General Editor:	David Jollands
Design Director:	Elwyn Blacker
Consultant Authors:	Roy Edwards
	Alan Hibbert
	Jim Hudson
	John Little
	John Mason
	Cleland McVeigh
	Peter Metcalfe
	Beverley Moody
	Patrick Moore
	Keith Porter
	Tim Pridgeon
	Derek Slack
	Ian Soden
	Tony Soper
	Alan Thomas
Research Editor:	Simon Jollands
Design and Production:	BLA Publishing Limited
	Michael Blacker
	Simon Blacker
	Margaret Hickey
	Graeme Little
	Alison Lawrenson
Artists:	Paul Doherty
	Hayward & Martin
	Dennis Knight
	Richard Lewis
	Steve Lings/Linden Artists
	Eric Thomas
	Rosie Vane-Wright

SCIENCE UNIVERSE SERIES

ENERGY, FORCES AND RESOURCES

**ARCO PUBLISHING, INC.
NEW YORK**

Acknowledgements

The publishers wish to thank the following organizations for their invaluable assistance in the preparation of this book.

British Telecom
Canon (UK)
Central Electricity Generating Board
Ford Motor Company
Kodak Museum
NASA
National Film Board of Canada
Philips International
Royal Greenwich Observatory
Royal Smeets Offset
Shell
Sony (UK)
Southern Positives and Negatives (SPAN)
Standard Telephones and Cables
United Nations Organization
US Information Service

Published by Arco Publishing, Inc.
215 Park Avenue South, New York, N.Y. 10003

© BLA Publishing Limited 1984

First published 1984

Library of Congress Cataloging in Publication Data

Main entry under title:

Energy, forces, and resources

(Science universe series; v. 4)

Includes index.
Summary: Briefly discusses energy sources such as food, sun, water, electricity, coal, gas, oil, nuclear power, as well as pollution and conservation.

1. Power resources – Juvenile literature. 2. Force and energy – Juvenile literature. 3. Power (Mechanics) – Juvenile literature. [1. Power resources. 2. Force and energy. 3. Power (Mechanics)]

I. Arco Publishing   II. Series

TJ163.23.E54   1984   621.042   83-26619
ISBN 0-668-06178-2

This book was designed and produced by
BLA Publishing Limited, Swan Court,
East Grinstead, Sussex, England.
A member of the Ling Kee Group
LONDON · HONG KONG · TAIPEI · NEW YORK · SINGAPORE

Phototypeset in Great Britain by
Southern Positives and Negatives (SPAN).
Color origination by Chris Willcock Reproductions
and Premier Graphics.
Printed and bound in The Netherlands by
Royal Smeets Offset BV, Weert.

Photographic credits

t = top   b = bottom   l = left   r = right   c = centre

Cover photographs: *tl, tr* E. A. Janes/NHPA; *bl, bc, br* ZEFA.

Title page: ZEFA.

4 ZEFA; 5*t* Shell; 5*b*, 6*l* ZEFA; 6*r* Central Electricity Generating Board; 7*t*, 7*c*, 7*b*, 8, 11*l*, 11*tr*, 11*cr*, 12*c* ZEFA; 12*b*, 13*t*, 13*bl*, 13*br* Bryan & Cherry Alexander; 14, 15*l*, 15*t*, 15*br*, 18, 19*cr*, 19*br* ZEFA; 23 Shell; 24, 25*t* ZEFA; 25*b* Space Frontiers; 26*l* Ann Ronan Picture Library; 27*lc* ZEFA; 27*br* Central Electricity Generating Board; 28*l* Mansell Collection; 28*cr* ZEFA; 29*r*, 29*b* Ford Motor Company; 32*t* Ann Ronan Picture Library; 32*b* Mansell Collection; 33*t*, 33*l* Royal Institution/Michael Holford; 33*b* Michael Holford; 34*t* Central Electricity Generating Board; 35*t* ZEFA; 38*l* Michael Holford; 38*tr* Mary Evans Picture Library; 40, 41*tr*, 41*c*, 41*br* Central Electricity Generating Board; 43*tl* Institute of Geological Sciences; 43*tr* ZEFA; 45*l*, 45*tr*, 45*br*, 46*bl*, 47*tr*, 48*bl*, 48*tr* Shell; 50*l* Sigmund Freud/Mary Evans Picture Library; 50*c* Mansell Collection; 51*t*, 51*b* Central Electricity Generating Board; 52*bl*, 52*tr*, 53*t*, 53*c*, 53*r*, 54, 55*bl*, 55*tr* ZEFA; 55*tl* Lacz Lemoine/NHPA; 56/57 ZEFA; 57*tr* Central Electricity Generating Board; 57*b* Philips; 58*t* Ann Ronan Picture Library; 58*b*, 59 ZEFA.

Conversion table for units

Length

1 nanometer (nm)	= 0.000001 millimeter	= 0.000000001 meter (one-billionth of a meter)
1 millimeter (mm)	= 0.1 centimeter	= 0.03937 inch
1 centimeter (cm)	= 10 millimeters	= 0.3937 inch
1 meter (m)	= 100 centimeters	= 39.37 inches
1 kilometer (km)	= 1000 meters	= 3280.8 feet = 0.621 mile

Area

1 square kilometer = 0.3861 square mile

Capacity

1 liter = 1.0567 quarts

Volume

1 cubic centimeter (cc) = 0.06102 cubic inch

Weight

1 kilogram (kg)	= 2.2 pounds	
1 metric ton	= 1000 kilograms	= 1.1 US tons

Contents

Introduction	4
What is energy?	6
Energy and the Sun	8
The importance of water	10
The environment	12
People on the move	14
Energy in the living world	16
Our daily food	18
How the body breaks down food	20
Plants and energy	22
Natural forces	24
Power from steam	26
The internal combustion engine	28
First encounters with electricity	30
The discovery of current electricity	32
The nature of electricity	34
Volta, Ampère and Ohm	36
Electricity and magnetism	38
Power stations and electricity supply	40
Coal and coal gas	42
Looking for oil	44
Offshore rigs	46
Oil refining and by-products	48
Nuclear power	50
Pollution of the environment	52
Conservation	54
Energy from Sun and wind	56
Energy from water and the Earth	58
Summary	60
Glossary	62
Index	64

NOTE TO THE READER: while you are reading this book you will notice that certain words appear in **bold type**. This is to indicate a word listed in the Glossary on page 62. This glossary gives brief explanations of words which may be new to you.

Introduction

WE ALL NEED ENERGY to do work. This comes to us from the food which we eat and which we use as our 'fuel'. In the same way an airplane uses fuel to carry a load of passengers over a long distance. Our energy needs are so immense that we sometimes wonder whether supplies will eventually run out. The supplies of energy are called **resources**.

This book is about energy and all the resources that we are able to use. However, energy supplies are becoming scarce and soon there will be new problems to face. We depend too much on coal, oil and natural gas as fuels for the modern world. These are called **fossil fuels**, and they were formed beneath the Earth's surface hundreds of millions of years ago.

Today, oil and gas are being used up too fast and all forms of energy are becoming more and more expensive. It is important to know how long these resources can last and how much they are going to cost in the future. We must also work out new ways of sharing resources more fairly between the developed and the developing countries of the world.

Human beings also need energy in the form of food. In fact, food and energy are closely related. If fuel prices rise, so do farming costs, and up goes the cost of food. The developing countries of the world cannot afford to import food and fuel. They may have to choose between using their scarce land for growing food, or using it to grow trees for fuel. Some countries now have a shortage of trees. As populations increase, the need for firewood in the villages also goes up. The forests are stripped bare and entire regions are left with no trees. These regions quickly become deserts.

In 1983 the world's population was just over 4 billion. By the year 2000 experts believe it will be over 6 billion. The average amount of fuel used in the world today has the same value as about 2.5 metric tons of coal per person each year. In some of the developing countries the average may only be a few hundred kilograms, but in the developed countries such as the USA it is over 12 metric tons. The developing countries need to use more energy to give their people the food and things they need.

During the ·1970s there was a simple connection between the amount of fuel used

Shell/Esso's Brent 'B' oil production platform in the North Sea is lashed by winds of more than 180 km per hour. The power of the wind and the 20-m waves seen here, could one day be harnessed, but this is unlikely to happen on a large scale until the world's oil supply runs low.

Cities consume massive amounts of energy, especially at night. Hong Kong, seen here, has to import most of its energy needs, including food and water. The costs are met by its advanced industries and its success in world trade.

nature. In this balance, the Earth receives energy from the Sun and radiates it back into space. We would not be creating new problems of **pollution** caused by such things as acid rain, oil slicks, or **radioactive** wastes. Some people say that the problems of energy shortage can be solved by nuclear power, but this would be very costly. Only very rich countries would be able to afford it.

By the end of this century the energy crisis will be with us. By then, fossil fuels will be in short supply, and very costly. Nuclear power cannot give us liquid fuels for cars and airplanes, but this can already be achieved by growing plants which produce alcohol fuels.

As we explore energy in all its forms, and the resources which are available, we will begin to see the problem. Our present way of doing things cannot last forever, and there is not very much time left for scientists and technologists to solve the problem. Meanwhile, we should conserve, or save, energy whenever we can, rather than waste it.

The burden of a large, expanding population results in problems with food supply. The energy crisis in India is one of the most serious in the world. Its difficult climate often leads to drought or floods. If a harvest is ruined, food can quickly run short for the population of about 600 million.

and the amount of **output**, or goods produced. To produce more, a country had to use more energy. Now there is not enough energy left for the developing countries of the world to use in the same way as the richer countries.

Fortunately, some of the large countries have seen this problem. The People's Republic of China, with one quarter of the world's population, has planned to increase its output four times by the year 2000, but it will use only twice its present amount of energy. To reach this goal it will develop **renewable** sources of energy.

What are renewable sources? They are sources which will never run out. They will be with us forever and there will never be a shortage. **Solar energy** is a term which we use to describe the energy coming direct from the Sun. Wind power and water power are other forms of renewable energy. Water power depends upon **evaporation** of seas and rivers caused by the Sun. Waves are formed because of the activity of the wind and the Sun.

If we use the energy of wind, waves or the Sun, we are keeping the natural balance of

What is energy?

WHAT IS ENERGY? Do you feel full of energy after breakfast, and lacking in energy at the end of a day's work? When you lie on the beach on a summer's day, are you taking in the Sun's energy? We use the word so much that it is easy to forget what it really means. The word energy comes from two Greek words meaning 'work within', or the ability to do work. Everything around us has some energy, although we may not always realize what kind of energy it has.

For example, one form of energy, **potential energy**, depends on how high something is above a fixed level. At a **hydroelectric** power station the water that is stored behind the dam has potential energy. When it comes rushing over the dam wall, the water then has **kinetic energy**, or energy of motion. The **turbines** in the power station convert this energy into electrical energy.

Modern ideas about the nature of energy were first thought out by Sir Isaac Newton in the seventeenth century. In his Laws of Motion he said that energy is neither created nor destroyed. It is changed into other forms of energy. If a power station burns coal to make electricity, all the energy in the coal can be measured. Some goes up the chimney in hot gases. Some goes to the cooling water and a large part of it becomes electricity. None of the energy is lost. It appears somewhere else as air at a slightly higher temperature, or as electrical energy.

When an airplane takes off and climbs, the energy in the fuel is changed. Some of the energy becomes kinetic energy – the airplane and the passengers are moving rapidly through the air. Some becomes potential energy – they are climbing higher and higher above the ground. Some heats up the surrounding air. After landing, the airplane needs to take on more fuel before it can do more work.

Although we have said that energy is neither created nor destroyed, some of it is nearly always lost because of the way we use it. The modern world uses a vast amount of electricity and we can no longer do without it. When fuels such as oil or coal are converted into electricity

A coal-fired power station. The stored energy in the coal is released by burning. Some of this energy escapes up the chimneys in the form of heat, but most of it is converted into electricity. The station consists of a coal bunker, boiler house and turbine hall.

This hydroelectric dam in Uganda shows an example of kinetic energy, or energy of motion. The torrents of water drive turbines which in turn produce electrical energy. Water stored behind the dam has potential energy. Once the sluice gates are opened this energy is released.

we use a lot of energy to do this. If 1 metric ton of coal or 0.6 metric ton of oil are burned to produce heat, each will produce as much heat as a 1 kilowatt electric heater would produce for 7350 hours. This is almost a year. However, if the coal or oil is burned in a power station to

The heavier an airplane is, the more energy it needs to power it. Before it takes off, calculations have to be made to ensure there is enough fuel for the airplane to reach its destination.

make electricity, energy is lost. First steam is produced. This drives a turbine to make or generate electricity. The electricity then has to be carried along cables and wires to reach our homes and factories. Our 1 metric ton of coal or 0.6 metric ton of oil would only keep the same heater working for 2000 hours instead of 7350 hours.

Energy helps to grow food, takes us from one place to another, keeps us warm and gives power to industry. Nearly all this energy comes to us from the fossil fuels, oil, gas and coal. Some

Our modern cities have massive energy needs. This picture of Tokyo, Japan includes many different energy uses – most of which depend on oil.

comes from **uranium**, the fuel used by nuclear power stations. You will see from the diagram that, if we continue to use these fuels at the same rate as today, the supplies will run out within a few years.

How do we deal with this problem if we are to maintain our way of life? We could look for more fossil fuels, but even this might not produce more than a few year's extra supplies. We could use less of these fuels, but this would only be possible for the developed countries, since the developing countries need more energy now to improve their living standards. Perhaps the only possible solution is to find new sources of energy which will not run out and which can be afforded by all countries.

A cyclist uses the energy his body takes from food to power his bicycle. Rather like the airplane, he performs best if he is not overweight and if his machine is as light as possible. The tires should be full of air, and the wheels well oiled to reduce friction.

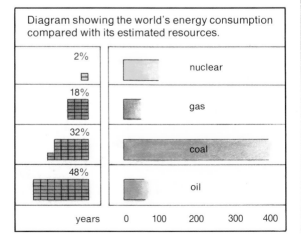

Diagram showing the world's energy consumption compared with its estimated resources.

2%		nuclear
18%		gas
32%		coal
48%		oil
years	0 100 200 300 400	

Energy and the Sun

THE SUN is the source of all our energy. We take it for granted. Every day we see it rise in the morning and set in the evening. It sends out energy in all directions. Scientists agree that it will last for thousands of millions of years. This is such an enormous time scale that we can think of the Sun as lasting for ever. It is about 1.4 million km in diameter and 150 million km from the Earth. However, in the universe, the Sun is very unimportant. It is only one of millions of stars in the Milky Way, which is itself only one of millions of **galaxies.**

The temperature at the center of the Sun is thought to be about 15,000,000°C. At this temperature, nuclear reactions are taking place the whole time, and these are the source of the Sun's energy. About 90 per cent of the Sun is made up of the gas **hydrogen.** Hydrogen is the lightest and simplest of the **elements.** The energy process inside the Sun is the **fusion** of hydrogen to make another gas, **helium.** This gives out very large amounts of energy. Fusion is the joining together of the atoms of light elements. It is the opposite process to **fission**, the splitting of heavy atoms. The latter is the process used in nuclear power stations.

At the surface of the Sun, the temperature is about 5800°C. A large part of the energy coming from the Sun is in the form of light. The greater the distance from the Sun, the less the light energy is concentrated. In the outer planets it is so low that life as we know it cannot exist. However, at the outside of the Earth's atmosphere, the total amount of energy received from the Sun's light is immense.

Some of this light energy, about 30 per cent, is reflected back into space. The balance provides direct heat at the Earth's surface, and also causes **evaporation** and rain. The amount of energy reaching the Earth from the Sun is thousands of times greater than that which we use from other sources, including wood. That is why scientists in countries all over the world are working on solar energy research.

Although the visible light from the Sun seems to us to be white, it is made up of different

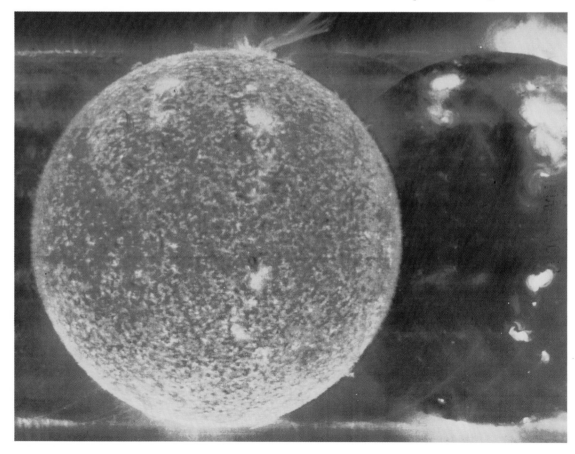

This false-color picture of the Sun was taken from Skylab during the missions of 1973 and 1974. A great deal was learned about the nature of the Sun that could not have been learned from previous earth-based observations.

About 30 per cent of light from the Sun is reflected off the Earth's atmosphere. The remainder passes through the atmosphere to the Earth's surface, causing heat, evaporation and rain.

colors which we see in a rainbow. A white surface reflects all these colors. On the other hand a black surface **absorbs** or takes in all the colors that land on it. Black is therefore the best 'color' for collecting the Sun's energy.

One of the ways in which we make use of the sun's heat is to place a surface to be heated under glass. This is known as the 'greenhouse' effect. Glass takes in the rays of the Sun rather like a mousetrap. The rays can pass through the glass, but once inside the solar 'mouse' cannot

countries, the sunlight is much weaker in the winter than in the summer. Although present during the hours of daylight, the Sun's radiation is never available for use at night.

For all these reasons and because some energy is always lost in the process of conversion, it is not easy to use solar radiation as a source of energy. However, solar energy can be stored. We have been using solar energy for hundreds of years. Every time we burn a log on an open fire we are using stored solar energy. Our fossil fuels, oil, coal and gas, also contain stored solar energy laid down millions of years ago.

Modern methods of storing such energy can convert the summer's sunshine into hot water.

The collecting and storing of solar energy using solar panels and a heat exchanger, which serves a hot water tank.

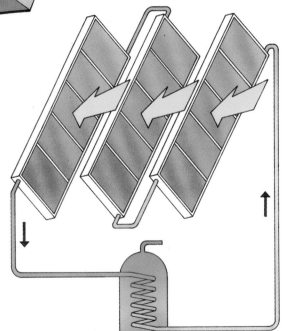

get out again. This is because the surfaces inside the glass are at low temperatures, perhaps less than 60°C.

The Sun does not shine all the time and there are differences in the amount of solar energy reaching the Earth's surface from one country to another. The axis of the Earth is tilted in relation to the Sun. This gives us the seasons, and a variation in the strength of sunlight between winter and summer. Apart from the tropical

This map of the world shows the variation in average solar radiation per annum. This is naturally greater in the tropical regions than in the temperate regions.

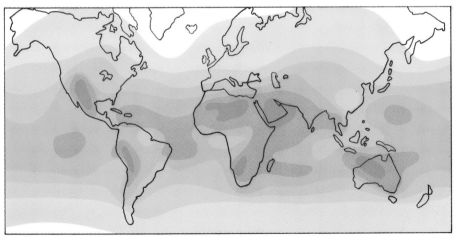

This water can be kept hot in large tanks which are **insulated** to reduce heat loss. The heat can then be drawn off in the winter as required.

Other storage methods include the use of special chemicals and solar ponds. These are shallow ponds containing a salt solution. In Israel a solar pond has been used to make electricity at midnight in mid-winter. This goes to prove that solar energy can be stored and used throughout the year.

The importance of water

THE OCEANS of the world cover nearly eighty per cent of the Earth's surface. Water is vital to all forms of life and nothing can live without it. For example, human beings need to drink at least 2.5 liters a day. We also need water for many other reasons. Think how many times a day you and your family use running water. People in Western countries use as much as 250 liters every day. Factories use even more. In developed countries they use more than half the total supply of fresh water.

Energy from the Sun heats the surface of the sea and causes **evaporation.** This means that the water from the sea turns into vapor, like steam. The sea is full of **mineral salts,** and these are left behind when evaporation takes place. The water vapor forms clouds and as the temperature falls it **condenses** back into water droplets which fall as rain or snow. The part that falls on land provides our fresh water supply. Animals, including human beings, use the water that collects in rivers and ponds. Plants use the rain that falls on the soil, taking it up through their roots.

Water is a **compound** of the two gases hydrogen and oxygen. There is nothing in pure water for plants and animals to feed on, but it carries in it dissolved chemicals that are important for life. For instance, when farmers fertilize their crops, the fertilizer has to be dissolved in water before the plant roots can **absorb** it.

Water can dissolve minerals, such as chalk, from rocks as it flows through them. This makes the water 'hard'. Hard water prevents soap from lathering properly, and it leaves a **deposit** in domestic water pipes which can slow down the flow of water. When this happens we sometimes say that the pipes are 'furred' up. The water which we call soft water is more pleasant to wash in and to drink.

A desert is a part of the land where very little rain falls at any time of the year. The plants and animals that live in deserts are able to store much of the water they take in. People live in deserts too. By digging deep holes they can sometimes find water underground and make wells. In the Sahara desert the people plant date palms around wells and grow crops. This kind of settlement is called an **oasis.**

All the great civilizations of the world have grown up near a plentiful supply of water. For example, the lives of the ancient Egyptians depended on the River Nile. Rain and the melting snow at the source of the Nile added so much water to the river that it flooded its banks in Egypt every summer. The flood left a layer of

The cycle of water. Heat from the Sun evaporates water from the sea and land. The water rises up into the air as vapor and forms clouds. The vapor condenses and falls to earth as rain, snow and hail. Some of it flows back into the sea.

fertile soil all over the river valley, and the Egyptians grew their crops there.

Moving water is full of energy. If you put your hand into a fast flowing stream you can feel the water trying to push your hand along. A water wheel was an early machine which made use of this energy. Each of the blades fixed around the wheel was pushed by the moving water to make the wheel turn.

Water wheels were often connected to large flat millstones. As the millstones were turned, the grain was ground between them to make flour. If the mill was built near a waterfall the

A water hole at a desert oasis. Note how the hole has been dug deep to reach the water. Oases are most numerous in artesian basins, where the ground level is very low and water is trapped in a layer of porous rock.

The force of a fast flowing river is very strong. Great skill is needed to control a canoe in such conditions, but the fuel is free and the sensation of power and speed a real thrill.

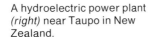

water could be channelled into a narrow stream which poured down on top of the wheel.

Water wheels use kinetic energy that comes from movement. Near the top of a high waterfall a river may move very slowly. However, when it reaches the edge it gushes over. Suddenly the water is moving very fast indeed. Water is full of potential energy. The stored energy in water can be used to generate electricity. In many parts of the world hydroelectric power stations have been built to make use of this energy. Dams are built across large rivers to hold back huge lakes of potential energy.

Instead of spilling over the top of the dam, the water can be led through a large pipe on to the blades of an enclosed water wheel, called a **turbine.** The blades are curved, rather like a ship's propeller, and they have to be extremely strong. The **mechanical energy** of the fast spinning turbine is turned into electricity.

Water, as well as being a vital resource for all living things, is also an important source of energy. Hydropower supplies 6 per cent of the world's energy needs.

A hydroelectric power plant *(right)* near Taupo in New Zealand.

Two types of water wheel *(below)*. An overshot wheel *(left)* may be sited at a drop in a river, or in a specially dug deep channel. The water is fed to the top of the wheel. An undershot wheel *(right)* draws its power from the flow of water striking its base. The wheel revolves in the opposite direction to the overshot wheel.

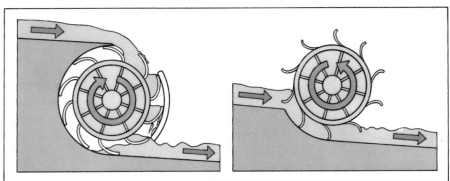

The environment

ALL LIVING THINGS are affected by their **environment**. Think of a tree in a forest. Its leaves touch those of the other trees around it, and spreading out in the sun they make a thick green covering over the ground. Down below it is cool and damp. As the falling leaves rot they help to make rich soil for the tree's roots. The forest, the cool earth, hot sun, rain and wind make up the tree's environment. Animals and smaller plants make their homes on the tree or in its shelter.

Your environment includes home, family, school, the weather, the view from your window, the noise of traffic, radio and television. The human environment is bigger than you may think. Cars, trains, ships or airplanes can now take us anywhere we want to go. Radio and television programs are sent all over the world. So we should really think of the whole Earth as the human environment.

The world population is now so big, and growing so rapidly, that we are in danger of changing and damaging the environment on which we depend for survival. The Earth cannot keep up with the speed at which human beings are changing it. If one or two trees are chopped down, new seedlings will soon take their place. In parts of South America whole forests have been cut down to make space for roads and for growing food crops. Unfortunately the soil that was held together by the tree roots has **eroded** or blown away. Only bare rock is left and nothing can grow.

In our world environment the materials we take from the Earth are not shared evenly between us all. Only one-third of the world population lives in the rich manufacturing countries, but it uses most of the world's resources. While people in Asia, Africa and South America have too little to eat, most people in other countries eat too much. Many of them have to diet or exercise to avoid getting fat.

All humans need food to eat, simple clothing to wear and some form of shelter. These, with water and air, are the basic necessities of life. We could survive with only these things if we had to, and some people do.

It is hard to imagine living in a desert, but plants, animals and even people have **adapted** themselves to all kinds of unfriendly environments. The Kalahari Desert in Southern Africa has no rain for five months in each year and the waterholes dry up. The bushmen who live there have to find drinking water by digging up the roots of plants that store water underground.

There are also cold deserts where it is just as difficult to survive. On the northern coasts of Canada, Greenland and Siberia there is snow for nine months of the year. The Eskimos live in this hostile environment of snow and ice. Because it is so hard to hunt food and keep warm, nothing the Eskimos take from their environment is wasted. They hunt seals and polar bears for food. They also use the skins for clothes and the fat for fuel. Since there are no trees they have to collect driftwood wherever they can during the short summer. They use this wood to

A group of bushmen and their wives in the Kalahari Desert.

An Eskimo in his kayak, made of skin over a wooden framework, paddles through the ice in North West Greenland.

Blocks of snow being used by an Eskimo to build an igloo in North West Greenland.

(below) Lapps follow their herd of reindeer on the spring migration in Norway.

A Lapp encampment (far right) on the Finmark Plain at sunset.

build houses, boats and sledges. In winter, when the Eskimos go hunting, the only building material is snow. They cut blocks of snow to build their dome-shaped igloos. Although the snow itself is freezing, warm air is trapped inside the igloo and this keeps the Eskimos warm.

Nomads are people who follow herds of animals as they **migrate**, or travel in search of food. They kill some of the animals for meat. But they protect the rest of the herd and make sure that the grassland is not destroyed by being overeaten before the herds move on. The Bedouin people in the North African desert keep sheep in this way.

In the far north of Europe the Lapps live in a cold and unfriendly land, herding reindeer. Like the Eskimos, the Lapps use every part of the animals they kill, for food, fuel, clothing and tents. They use the animal bones to make tools. The Lapps are always on the move and their temporary settlements leave little evidence on the land.

Bushmen, Eskimos, Lapps and other nomads are not a threat to their environment. In fact, they help to preserve it. There are, however, very few communities of this type in the world. Our environment is in danger because we do not care for it enough. The Eskimos are forced into making the best use of their resources while, at the moment, we are not. We have to think now of new ways to live in our environment without destroying it.

People on the move

THE NUMBER OF PEOPLE in the world is increasing at a rapid rate. Every minute about 250 babies are born somewhere in the world. Since 1900 the world population has doubled. With so many people in the world, our towns and cities cannot house and feed them all. Food has to be moved to places where people crowd together. As the cities expand, people have to move out into new areas and build new towns and homes.

Our early ancestors were hunters, and they followed herds of wild animals. They depended on these herds for survival and had to live in temporary camps, moving on when the herds moved. In some regions of the world food was plentiful and humans lived in small **settlements.** Before people could settle in an area there had to be plenty of food, water and shelter.

As man developed tools and learned to grow crops, these settlements became permanent and grew into villages. Each village grew all the food and made all the tools it needed. We call this **self-sufficiency.** As civilization developed language, and the desire for a wide variety of material goods, each village needed many skills to remain self-sufficient. Not all villages would have the moist clay needed to make good pots, or the iron ore needed for tools. One village would exchange arrowheads for cooking pots from another village. This was the start of movement of material and people between villages.

The villages gently broke away from self-sufficiency by sending their goods to a nearby center, or market. These 'centers' were usually small villages built where roads met. Eventually they grew into towns.

As civilization developed, nations became expert in making and trading in certain goods. Just as the villages sent their goods to market, products from all over the world were sent to the centers of great empires. Slaves and goods were shipped from all parts of the known world to Italy during the Roman period.

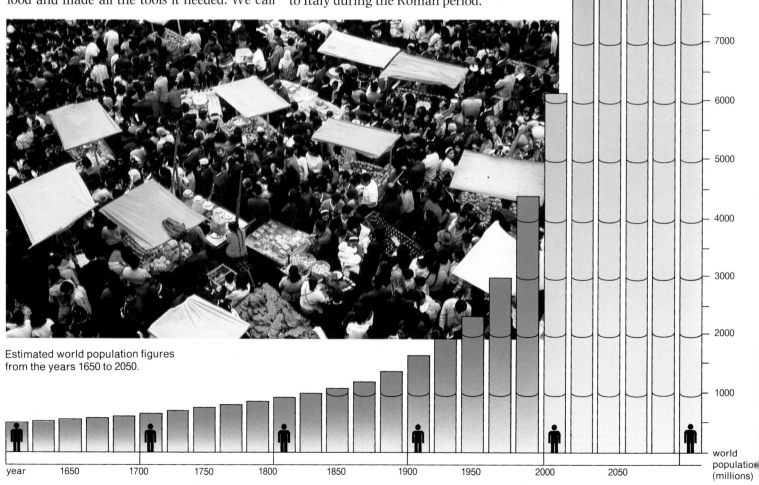

Estimated world population figures
from the years 1650 to 2050.

year 1650 1700 1750 1800 1850 1900 1950 2000 2050

world population (millions)

13000
12000
11000
10000
9000
8000
7000
6000
5000
4000
3000
2000
1000

The nature of goods and produce always decided what type of transport would be used to move them to market. Fresh meat, vegetables and fish need to reach market as quickly as possible. Bundles of cloth and cooking pots can be transported slowly, hence cheaply. To move something quickly uses a lot of energy. A parcel can reach the furthest parts of the Earth in a few days by airmail. If we wanted to send it cheaply it could go by sea-mail but it would take several weeks.

More energy is needed to move something quickly. Nature designed human beings to move most cheaply, or **efficiently**, at walking speed. Our muscles and body allow us to walk for long distances without too much discomfort. Nature has also given us the ability to run swiftly when we need to. When we run, our bodies switch over to a special way of producing energy. The body does this by producing energy in the absence of **oxygen**. We pay for this after running by breathing quickly and deeply. We do this to put back the oxygen and to get rid of heat.

Human beings have tried to improve upon nature's way by inventing faster ways of travelling. All forms of transport need fuel, even gliders need to be towed into the air. The ideal form of transport and the speed at which we travel will depend upon the need for that travel. The machines that transport us have a speed at

Modern jet airliners provide the fastest transport links between countries and continents. The wide-bodied Boeing 747 Jumbo Jet can fly nearly five hundred passengers at 800 km per hour. With a full load, each passenger will be transported over 100 km on 4 liters of fuel.

(below, left) Where labor is cheap and fuel is costly, the best engine is often man himself. This load in Canton, China, may take some time to reach its destination, but it can be carried about 5 km for the cost of two square meals.

The *Queen Elizabeth 2* cruise liner *(below, right)* provides luxury travel but at high cost. It takes a passenger about 20 km on 4 liters of fuel, at about 45 km per hour.

which best use is made of the fuel. This is called the **optimum** or **cruising speed**. We measure how cheap it is to run a car by checking how many kilometers it will travel on 1 liter of gasoline. Many modern cars travel most economically at around 80 km per hour.

Perhaps the best form of transport yet invented is the bicycle. This is used all over the world as a cheap and clean form of transport. We use about as much energy pedaling a bicycle as we do when we walk, but we can move as fast as we can run.

Energy in·the living world

THE LIVING WORLD is a phrase that is used in biology. This living world, or **biosphere**, as it is called by scientists, is a world which stretches from about 10,000 m below sea level at the bottom of the deepest ocean to a height of 9000 m above sea level at the top of the highest mountains. Almost all the Earth's surface is part of the biosphere.

It is difficult to think of energy in the world as a whole. Instead, we use a much smaller unit of life that we call an **ecosystem.** This is a small group of living things that can exist on their own without outside help. An ecosystem can be as small as a garden rain barrel or as large as a forest. One thing that all ecosystems have in common is the need for the energy in sunlight to make them 'work'.

The first and most important parts of any ecosystem are the green plants. These trap the energy in sunlight and use it to make food. They need only water, air, mineral salts from the soil and sunlight. For this reason we call them producers or **autotrophs.** All other **organisms** must use food already produced by green plants and these are called consumers or **heterotrophs.**

Animals that eat plants are called **herbivores.**

All herbivores have natural enemies which do not eat plants. These are called meat-eaters or **carnivores.**

Let us think how all this fits in with a simple ecosystem such as a pond. In the pond are water plants which make their own food using the sun's energy. These plants may be eaten by water snails which are the herbivores. In turn the snails may be eaten by fish such as perch. The perch is the carnivore because it eats only other animals. This is a simple example of what we call a **food chain.**

The snail would eat many kinds of water plant and the perch would eat many other animals in the pond. This is a **food web** which connects all the animals and plants in the pond. Food webs can be very complicated because of the many types of animal and plant which live in even the simplest ecosystem.

In our simple food chain, the energy of sunlight 'locked' into food by the plants, is passed along the chain. The snail will depend upon the plant leaves and the perch will depend upon snail 'meat'. At each stage, only part of the original energy in sunlight will be passed on in the food. By far the greatest part will have been

The picture illustrates a natural community of living things. We call this an ecosystem. It consists of living things including plants and a non-living environment of earth and water. The ecosystem depends upon the energy it receives from the Sun, and on a continual supply of water.

used up by the plant or animal in keeping itself alive. The plant needs energy to carry out its daily activities and also for new growth. The snail needs energy for the same reason and also to move. The perch needs energy to make muscles work and keep active. Much of this energy will be lost as heat into the pond or will be turned into waste, as it is passed along the food chain.

The waste from all organisms will fall to the bottom of the pond. Here it will be broken down into water, gases and mineral salts by **decomposers.** Any energy that did remain in the waste will finally be lost as heat. This is the end of the line for energy, it is not **recycled**.

If we could take all the organisms out of the pond and count them we would find that their numbers would fit into a pyramid shape. Most of the organisms would be plants (producers).

These would be at the base of the pyramid. A smaller number would be herbivores and the remaining few at the top of the pyramid would be carnivores. If we worked out how much energy each organism was worth we would make a similar pyramid. Plants, at the bottom of the pyramid, would have by far the greatest total energy store in the pond and the animals at the top the least.

In going along the food chain, energy is wasted in each stage as the plants and animals use most of it to keep alive. If man could live by eating just plants he would save the energy that would otherwise be lost in the normal food chain. Farmers would get much more food, in energy terms, from a field of wheat than from keeping cows in the field. However, humans need a varied diet which includes the protein found in meat.

In any natural community a food chain is a simple series of living things in which each member is linked by food needs. The first link is a plant. This is eaten by a herbivore (plant-eater). The herbivore is in turn eaten by a carnivore (flesh-eater).

The population of a natural community can be shown as a pyramid diagram. There are always many more green plants at the bottom of the pyramid than there are carnivores at the top.

An ecosystem contains many food chains and these combine to make a food web. Carnivores and herbivores vary their diet when there is a lack of their favorite food.

Our daily food

THE HUMAN BODY can be compared to an engine. Like all engines, the body needs fuel in the form of food, providing the body with energy for warmth and work. Food also provides material for growth and repair of the body. Contained in food are small amounts of other materials which keep the body in good health. All these substances are called **nutrients** and they are needed in our daily food or diet. A frequent intake of water is also essential.

Scientists are able to measure the energy value of food by burning small samples in a special machine called a **calorimeter.** This measures how much heat is produced by the burning food. The units we give to this measurement are called **calories** or **joules.** One calorie is equal to 4.2 joules. In order to keep fit and healthy, human beings need to take in a certain number of calories each day. A twelve-year-old boy needs about 3000 calories, while a girl of the same age needs 2500 calories each day. If we eat more than we need, the extra food is changed into fat and stored.

The nutrients which produce energy and heat are the **carbohydrates** and **fats.** Both are broken down by the body and are used to help us to do work or to produce heat. The body-building nutrients are called **proteins.** The

There are many people in the world who do not get enough food and there are others who eat too much. The amount of energy in food is measured in calories or joules. This chart shows how the caloric needs of people of various ages change through life for each kilogram of body weight.

birth · 6 months · 1 year · 2 years · 4 years · 8 years · 12 years · 18 years · 70 years

proteins in food have to be broken down into smaller substances called **amino acids** before we can use them. These are then used to build new tissues, the substances of which our bodies are made.

Other important nutrients are the **vitamins, mineral salts** and **roughage.** There are many different vitamins and each of them is needed in tiny amounts. Every vitamin does a different job and the absence of only one can seriously affect our health. Long ago, before we realized the importance of vitamins, seamen used to die of a disease called scurvy on long sea voyages. Scurvy is caused by a lack of vitamin C, which is plentiful in oranges and lemons. Seamen found that by eating these fruits they could prevent scurvy.

Mineral salts are important for making bones and teeth. Without simple salts our muscles and nerves do not work well. If we do not take in enough salt we may get cramps in our muscles.

Another part of food is the fibrous type of carbohydrate that we cannot digest. This is called roughage and is thought to help in keeping the passages of our **intestines** open.

Not all foods contain every nutrient needed by our bodies. Most foods are rich in one or perhaps two nutrients. Meat, fish, cheese and eggs are rich in protein. Foods with flour or sugar in them are rich in carbohydrates. The fat-rich foods include butter, milk, nuts and some fish. Fresh fruit and vegetables contain all the vitamins and mineral salts we need, and they also give us most of our roughage. Each day you should try to eat food from each of the above groups. This is what we call a balanced diet and it will give you all the nutrients needed for healthy growth.

People throughout the world have different eating habits. North Americans and Europeans eat plenty of meat. Chinese and other Asians eat large amounts of rice and vegetables. The Japanese, and to a lesser extent the Russians, are fish-eaters. Each way of life depends upon the amount and nature of land available, and how many people have to be fed from that land. The Chinese and the Indians, with their

The human body needs a variety of foods to work and grow properly. To be strong and healthy a young adult needs a balanced diet similar to this one.

Cereals are the world's basic food. Wheat is grown in most non-tropical or temperate parts of the world. Wheat grains are ground to make flour for bread.

Food varies from one country to another. All human beings need certain nutrients; carbohydrates such as bread and sugar; fats in dairy products and meat; protein in meat, eggs and fish, and roughage in vegetables.

Beef cattle in Colorado. Beef cattle are reared for meat on grasslands and prairies. The herds consist of males, known as bullocks or steers, whose reproductive organs have been removed.

large populations, have to provide their food from poor land, in a difficult climate. They do this by using the land to its best advantage.

If we think of food in terms of a food chain, we know that energy is lost at each stage going up the chain. In fact, a farmer can supply ten times more food from a field by growing wheat on it than he can by using the field for rearing cattle as food. In Europe and North America there is plenty of farm land compared with the population. Perhaps it is for this reason that we allow ourselves to waste energy from the Sun by eating animals that eat plants.

How the body breaks down food

ALL FOODS have to be broken down into very simple forms before the human body can use them. These nutrients can only be used by the body after they have been dissolved in the blood. The breakdown, or **digestion,** is done by our digestive system.

The digestion of food begins in the mouth. There is a good reason for chewing your food before you swallow it. As you chew, the food is broken down into small pieces and mixed with **saliva.** The saliva is produced by glands in the mouth and contains a substance which helps to break up food. This substance is known as an **enzyme.** Throughout the digestive system there are many different enzymes, each breaking down a particular part of the food. The enzyme in saliva changes **starch**, found in potatoes, bread and cakes, into sugar. Try chewing a lump of bread for a while and it will soon taste sweet as the enzyme works.

When it has been chewed and mixed with saliva, the food can be swallowed. The saliva helps the food to pass down the tube leading to the stomach by acting as a lubricant, just as oil in a car engine helps it to run smoothly. The food is squeezed down the tube, or **esophagus**, by the action of muscles. At the bottom of the esophagus the food reaches the stomach.

The stomach is a muscular bag which acts like a cement mixer. It churns up food with yet more enzymes and a strong acid. The digestion of proteins begins in the stomach. Food can stay in the stomach for several hours, but eventually it leaves and passes down a long tube called the **intestine.** The intestine is made of two main parts, the small intestine and the large intestine. This is not because one is longer than the other but because one is wider.

In this drawing of the human digestive system, some parts, such as the windpipe to the lungs, have been cut away so that you can see the digestive organs clearly. Food is prevented from entering the windpipe by a 'trap-door' known as the epiglottis. The liver is a large gland near the stomach. It has many duties, storing minerals and vitamins, and maintaining the chemical balance of the body. It also makes bile which is stored in the gall bladder (shown slightly out of place in the illustration, for clarity), until needed for the digestion of fats. The pancreas, a gland near the duodenum, supplies enzymes. Also shown in the picture is the appendix, which seems to serve no purpose in human beings.

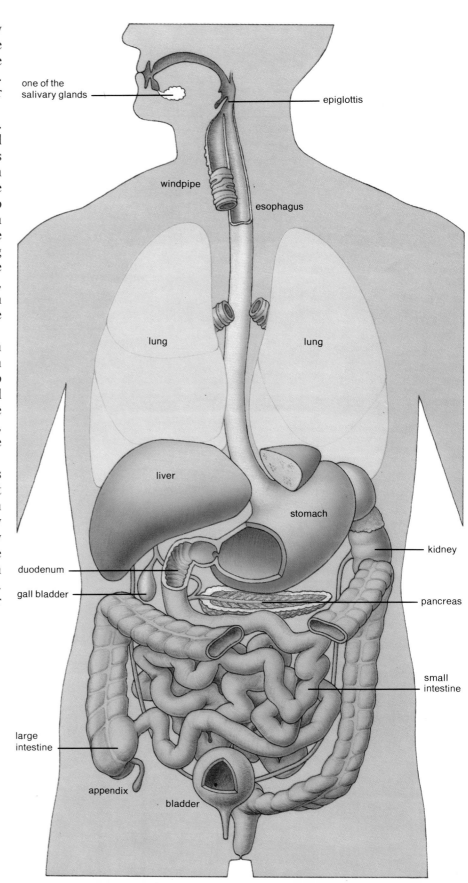

one of the
salivary glands

epiglottis

windpipe

esophagus

lung

lung

liver

stomach

kidney

duodenum

gall bladder

pancreas

large
intestine

small
intestine

appendix

bladder

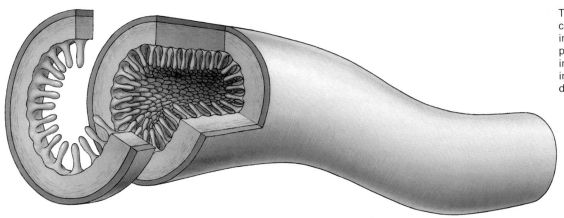

This diagram shows an enlarged cross-section of the small intestine. The small hair-like projections (known as villi) increase the surface area of the intestine for absorbing and digesting food as it passes by.

The small intestine is very nearly 7 m in length. The first 30 cm is usually called the **duodenum**, and here more enzymes are added to the food, and more digestion takes place. At this point the food has become a creamy, thick liquid. It contains a rich supply of sugars from digested carbohydrates, and some amino acids from broken down proteins. Digestion is completed in the small intestine. The wall of the small intestine can be compared with a piece of carpet rolled up with all the pile inside. This helps to give the small intestine a very large surface area so that the digested food can be absorbed easily in the fine blood vessels of the intestine's inner wall.

After most of the nutrients have been absorbed, the waste food passes along the large intestine. This is only 1.5 m long and is there to soak up all the water from the waste. The unused, or indigestible, parts of food pass right through the large intestine. They leave our body from the last section of the large intestine known as the **rectum**. The correct word for this waste is **feces.**

The energy-giving foods, fats and carbohydrates, have been broken down by the time they reach the small intestine. There they pass into the bloodstream and are taken all around the body. Some sugars can be used immediately to provide energy. Excess sugars are passed to the **liver** for storage. The excess broken down fats are stored under the skin and around organs such as the kidneys and the heart.

The products of digestion are used by the body for energy and for building tissue. Once the energy has been taken from food, the waste that is left has to be taken out of the body. When the body 'burns' up sugars it produces carbon dioxide and water. The carbon dioxide is carried in our blood to the **lungs**, where it is exchanged for oxygen. Water may not seem to be a waste product, but too much can upset our life systems.

Many other waste products from the breakdown and wear of cells are dissolved in the water of our blood. The blood passes these unwanted materials through our **kidneys** which act as filters. Here the waste is removed, along with water, and it is passed out of our bodies as urine. Thus the kidneys are the last step in the passage of energy through our bodies, by getting rid of the waste.

epiglottis

esophagus – food swallowed in small lumps

one of the salivary glands

stomach – digestion starts, acid and enzymes added

duodenum

gall bladder

pancreas

small intestine – digestion continued and completed

large intestine – water absorbed

rectum–feces formed

anus – waste removed

The alimentary canal

Plants and energy

WITHOUT GREEN PLANTS there could be no life on earth. Our world receives energy from the Sun and plants trap this energy and use it to make food. All animals depend upon the food produced by green plants.

Not all the energy in sunlight is converted into food by green plants. For every one hundred units of energy contained in sunlight reaching the plant, between one and five units only will be made into food. Over half of the sunlight will be reflected by the leaves of the plant and the rest will be absorbed. Of the one to five units that are used in making food only half will be passed on as food to animals. This is because the plant itself has to use some of this food to keep itself alive.

All green plants obtain their food by a process called **photosynthesis.** In this process, water and a gas from the air called **carbon dioxide,** are joined together to form a sugar called **glucose.** During photosynthesis **oxygen,** another gas found in air, is released as a waste product. The glucose formed during photosynthesis is the basic unit of food on which all life depends. Without sunlight photosynthesis could not occur. The sunlight provides the energy that is needed to join water and carbon dioxide. This energy is then stored in the glucose produced ready to be used by plants or other animals.

In addition to sunlight, air and water, plants also need mineral salts from the soil. These are brought together in the leaves where photosynthesis takes place. Air and sunlight are plentiful and the leaves are surrounded by these. Water and mineral salts, however, have to be taken from the soil by the roots and then to the leaves. Plant roots are very important as collectors of water and dissolved mineral salts. They also act as an anchor for the plant and as a place where food can be stored.

The stem of the plant is the pathway along which water and mineral salts pass to reach the leaves. At the same time, sugars made in the leaves need to move down the stem to be stored in the roots. The stem has separate tubes to act as a one way system. The tubes that carry water and minerals are called **xylem** and the ones that carry sugars are called **phloem.** To do their job, plant stems need to be tough yet flexible. The building material for the stems is a substance called **cellulose.**

All organisms get energy from glucose by using oxygen to reverse what happens in photosynthesis. This process is called **respiration.** It produces carbon dioxide and water from glucose and oxygen. Plants need energy to move water and mineral salts up from the soil. This energy cannot come directly from sun-

In photosynthesis *(left)* carbon compounds such as glucose (yellow) are made by the plant in the presence of sunlight. Water and minerals (dark blue) and carbon dioxide (red) are taken in. Oxygen (pale blue) is given off. The reverse process of respiration *(right)* takes place at night. Oxygen is taken in but carbon dioxide is given off.

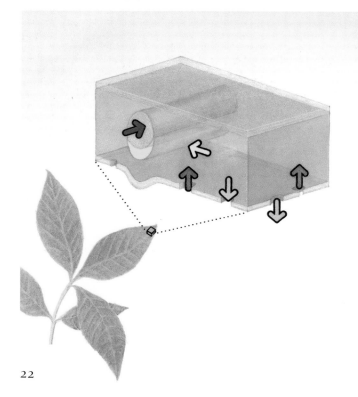

Pulling rice seedlings ready for transplanting in a paddy field in Brunei. Over half the world's population feed mainly on rice, particularly in Asia. When they are about 20 cm tall the seedlings have to be transplanted by hand.

For a plant to survive it needs food, water and minerals. It also needs a two-way transport system to carry these. The main highway is the stem. Food is made in the leaves by the action of sunlight on a pigment known as chlorophyll. It is carried in the mainly 'down' part of the transport system. It goes to the parts of the plant where it is used or stored, in long tubes called phloem cells. The 'up' part of the system carries water and minerals from the earth in long cells called xylem. Xylem and phloem cells in stems are scattered in bundles or in rings. In roots, xylems are concentrated in the center.

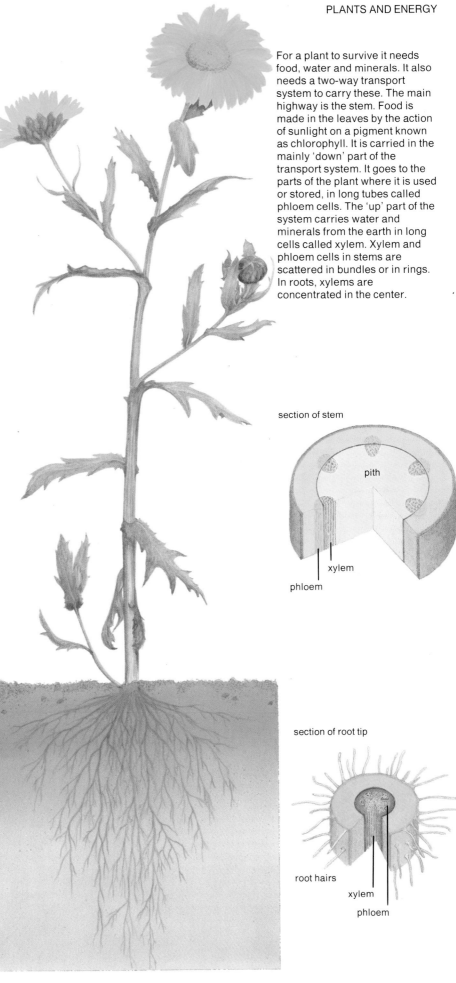

section of stem

pith

xylem

phloem

section of root tip

root hairs

xylem

phloem

light. Instead, the plant first has to make glucose and 'burn' this up in respiration to get back energy. This seems wasteful but energy has to be in the right form to be useful. The glucose contains **chemical energy** – sunlight contains **light energy.**

Plants need a good supply of water and minerals to grow well. Sunlight and air are plentiful. The soil provides the plant with water and minerals and the amount of each in the soil affects its **fertility.** The farmer tries to improve his crops by using fertilizers. In many parts of the world plants will not grow because of lack of rain. Lack of either water or mineral salts can stop plants growing, even when there is plenty of energy available in sunlight.

One of the greatest steps forward in our development was when human beings began to grow **cereal** crops. These include rice, oats, wheat, barley and maize. These crops produce the largest amount of food possible in a given area. The fruits of cereals, or grain, are rich in protein, vitamins and carbohydrates. When people learned to grow crops as food, they were able to give up hunting animals and were able to settle down in villages and hamlets.

Natural forces

WE ALL KNOW what a force is, but we probably find it difficult to explain this clearly in words. A force is not something you can see, you can only see the effects, that is to say what the force is doing. The force of the wind and of moving water can be used to do work for us, but we cannot see the force itself. To find out more about natural forces, we need to look at some of the effects.

Imagine a car that has run out of gas standing on a level road. It will stay where it is unless someone moves it. We will have to apply a force to make the car move. A force is anything that can start an object moving. Suppose we give the car a good push – a pushing force. The car will roll along the road. It will either slow down and stop, or it may hit another car and stop. If it slows down and stops, a **friction** force is what stops it. If it hits the other car another pushing force is what stops it. It could even bounce off the other car and change direction. A force is also anything that can cause an object to stop moving, or change its direction when it is moving.

The real test for a force is to decide whether it can make something start or stop moving. A force can also alter shapes. Pushing forces cause **compression**. We use compression to squeeze juice from an orange. The pulling forces produce **tension**. Tension in a bowstring can certainly make an arrow move forward.

What happens if an object starts moving and there is no force to stop it? A car will not slide or skid very far on a road, because friction slows it down and stops it. If there is water or ice on the road, the car can slide a long way because friction is very much less. With no friction at all the car would go on sliding forever, provided there was nothing in the way to stop it. Our Earth has revolved around the Sun for hundreds of millions of years because there are almost no forces to slow it down.

If you are sitting on a chair while you are reading this book, are there any forces in action? You are not moving, you are sitting quite still. A force called gravity is pulling you downwards and the chair itself is pushing you upwards. These two forces balance each other and this explains why there is no movement. Forces like this are said to be in **equilibrium**. An airplane in flight has two pairs of forces in equilibrium. Its weight is balanced by the lift forces from the wings and its drag is balanced by the thrust of the engine.

An airplane is kept flying by a balance of forces. The wings need to be of a special shape called an airfoil. As the air rushes over the upper surface of the wing a lifting force takes place. This is sufficient to overcome the force of gravity. The force of the engine, the thrust, drives the plane forwards. This thrust has to overcome the drag caused by air resistance and friction.

Bobsledding at St Moritz, Switzerland. Very high speeds of 140 km per hour or more can be achieved. This is due to the force of gravity and the little friction between the metal blades and the ice.

Space shuttle Columbia at blast-off. A rocket needs an immense amount of thrust to push itself and its load away from the Earth's gravity. The upwards movement, the lift-off, is achieved by the jet of burning gases going in the opposite direction. To withstand the G forces the astronauts have to lie flat on take-off. To escape from the Earth's gravity any spacecraft has to reach a speed of over 28,000 km per hour.

Gravity is a force that plays such an important part in our everyday life that we take it for granted. We know that if we drop a stone it will fall vertically to the ground. It is attracted to the center of the Earth and would go on falling in that direction if it could.

What is most puzzling to us is that gravity can act at a distance. No force is visible between the stone and the ground, nor have we pushed or pulled the stone. We have just let it fall. It is pulled by gravitational attraction towards the Earth's center.

This force of gravity gives weight to all objects, including ourselves. There are a few occasions when we notice a change in weight. Just imagine yourself in an express elevator on the ground floor of a skyscraper or apartment building. As the elevator speeds upwards you have a feeling of increased weight. This is due to gravity plus the increased pushing force beneath your feet. As the elevator comes down again the pushing force beneath your feet is reduced and you feel lighter.

Astronauts in space experience variations in weight as part of their job. When a rocket is launched it accelerates to a speed of about 28,000 km per hour. It reaches this speed within twelve minutes of blast-off. As a result, the astronauts experience.G (gravity) forces of up to five times their body weight.

When the spacecraft goes into orbit the pull of the Earth's gravity is cancelled out by other forces. The astronauts, the spacecraft and everything inside it become weightless. We sometimes call this zero gravity. The spacecraft needs no fuel to keep it in motion. There is no atmosphere, so there is no friction to stop it moving. It is in free fall.

If you were to stand on scales in an elevator, your weight would appear to increase, going up, and to decrease as the elevator goes downwards.

Space walk carried out on the sixth shuttle mission. The astronauts are weightless. The force of gravity has been cancelled out by other forces; they are in 'free fall'.

25

Power from steam

NOWADAYS when there is work to be done we often have a machine to help us. In prehistoric times human muscles were the only source of power. Later, animals such as the horse and the ox were tamed. Animals were able to carry goods, pull a plough or even drive a simple machine for pumping water. The natural power of the wind and flowing water were used to grind corn. However, muscles get tired and the wind did not always blow when people wanted to grind corn. Fast flowing water was not always found where the power was needed.

As civilization developed, the need for a powerful engine developed too. Engines are machines that make things move. They supply power but they have to have fuel to make them work. The engine which gave a completely new source of power was the steam engine. It was largely responsible for the Industrial Revolution.

The way power is obtained from steam is quite easy to understand. When water is heated in a boiler above its boiling point of 100°C it turns into steam. Just 1 cc of water will **expand** when boiled into 1700 cc of steam. In a closed **cylinder**, a **piston** can be made to move as a result of this expansion. When the cylinder is cooled, the steam **condenses** back into water. Since the water only takes up 1/1700th of the space in the cylinder, a partial **vacuum** is formed and the piston is sucked down again.

It is thought that the first person to make a steam engine was a Greek mathematician called Hero of Alexandria. He lived in the first or second century BC and invented a number of mechanical devices. His 'Ball of Aeolus' used jets of steam to spin the ball.

By the end of the seventeenth century in Europe, much of the easily mined coal near the surface had been used up. Mines had become so deep that flooding became a serious problem.

(opposite) Steam power was a great invention which transformed the world. The idea of the piston and the cylinder was possibly just as important since it made steam power work. In a simple steam engine steam is raised in a boiler and passed into a cylinder. As this happens the piston is forced back and a crankshaft turns a flywheel.

Thomas Newcomen's steam engine was also used to pump water out of coal mines. It was used at a coal mine near Dudley Castle in Staffordshire, England. We can imagine the noise and clatter near one of Newcomen's engines as the beam clanked up and down sixteen times a minute, spilling out water every time it did so.

Thomas Savery's first steam engine *(right)*, was known as 'the miner's friend'. The engine itself had to operate down the mine since it could not pump water from a depth greater than 6 m.

The Engine to raise Water by Fire

Printed for T. Hinton at the Kings Arms St Pauls Church Yard. 1747.

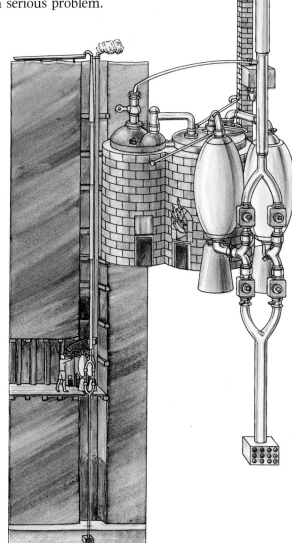

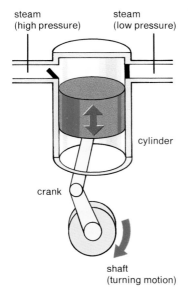

steam
(high pressure)

steam
(low pressure)

cylinder

crank

shaft
(turning motion)

Underground tunnels were dug to try and drain off the water. Various pumping machines worked by horses or men were tried but without success. There was a desperate need for more powerful pumps.

The first working steam engine was built in 1698 by Thomas Savery, a military engineer. In Savery's engine, steam was piped into a container and then shut off by a tap. Cold water poured over the container cooled it, and the steam inside condensed back into water. This caused a partial vacuum inside the container which sucked up water from the mine. Steam was then let in again and this forced the water up a pipe and out of the mine. Savery's engine used two containers.

Nobody knows for certain whether Savery's pump was actually used in a mine, but it could certainly pump faster than a hand pump. It was very wasteful of fuel although coal from the mine was cheap. Its great disadvantage was that it could not suck up water from a depth greater than 6 m. This meant that the pump and the boiler had to be underground for the pump to do useful work.

Thomas Newcomen was an ironmonger who lived in Cornwall, England. He often visited the Cornish tin mines because of his job and knew of the problems of flooding. Newcomen did experiments with steam engines for fifteen years, and in 1712 he built his first full-size engine. His engine was an **atmospheric engine**, so called because the piston was forced down again by air pressure.

His engine worked quite simply. Steam was let into the cylinder and up went the piston. The beam rocked to the left helped by the weight of the pump rod. Now a jet of cold water quickly cooled the steam inside the cylinder. The steam condensed into water and made a partial vacuum.

The vacuum was beneath the piston so the air pressure on top of the piston pushed it down. The beam rocked to the right pulling up the pump rod, and each time the pump rod was pulled up about 50 liters of water were pumped out of the mine. His engine was able to pump at the rate of 600 liters per minute.

The Newcomen engines used a lot of fuel, but they were the best engines available. Their design hardly changed over the next fifty years. In 1722 the first Newcomen engine was exported, to Czechoslovakia. By 1775 there were over 160 in use, and the biggest of these had a piston nearly 2 m in diameter.

Steam engines in which a piston moves up and down in a cylinder are no longer used. However, high pressure steam is widely used to drive the blades of a turbine. Most ships and electricity power stations use steam turbines, even though they burn oil or coal to create the high pressure steam.

Standard class steam locomotive 'Britannia' here seen in working order. Steam locomotives such as this one were a familiar sight on British railways fifty years ago.

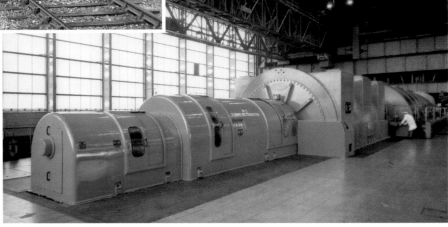

A modern turbo-generator used in a coal-fired power station. It is the power of high pressure steam which even now drives the turbines in electricity power stations. This applies whether they are coal, oil or nuclear-powered.

The internal combustion engine

THESE DAYS we think of the automobile as a necessity. However, at one time, it was a luxury which only a few people could afford. Although the automobile gives us the freedom to go where we like, it is an expensive form of travel. Transport is the word we use to describe the movement of people and goods from place to place, whether by land, sea or air. We know that the world's oil reserves are running out, and oil can only get more and more costly as the years go by. Yet oil provides over 90 per cent of all the fuel needed to keep the world's transport on the move.

Different types of transport use different amounts of fuel. The world's railways use the least, ships come next, then airliners. Road vehicles, however, use most fuel. They consume over 75 per cent of all the energy used by the world's transport. This energy comes mainly from gasoline or diesel fuel. Of all road vehicles, the private car uses the most energy for what it does. A bus can carry fifty or more people on only twice the amount of fuel used by a car carrying four or five people the same distance. Frequently you will see a car carrying only one or two people.

Most large cities of the world have severe traffic congestion, particularly when people are driving to and from work. One way a nation can save energy is to restrict the use of private cars in cities, and encourage the use of public transport.

The root of the problem lies in the way fuel is converted into mechanical energy. Nearly all road vehicles are powered by the **internal combustion** engine. As we have learned, a steam engine burns its fuel in a boiler which is separate from the engine. The internal combustion engine uses pistons and cylinders, like the steam engine, but it burns the fuel inside the cylinder.

In the early part of the nineteenth century, steam driven 'cars' were tried out on the roads, but they were a failure. The coal or wood used as fuel was too bulky and the search went on for a light, efficient engine for use in road transport. The breakthrough was made in the 1870s by a German engineer named Nikolaus August Otto. He used ordinary **town gas** for fuel and developed a new idea which is still used in our present-day cars.

In a four-stroke gasoline engine, each up and down movement of the piston is known as a 'stroke'. There are four strokes to complete one cycle. At the first stroke a mixture of fuel and air is sucked into the cylinder, above the piston. At the second stroke the mixture is squashed or compressed. At the third stroke the compressed mixture is ignited and there is an explosion. This forces the piston down violently. It is the power stroke. At the fourth stroke the piston rises again and forces the exhaust gases out of the cylinder. This happens over and over again, as long as the engine is running.

His engine differed from other gas engines of the time in one special way. It **compressed,** or pressed down the gas tightly in the cylinder, before it was burned, or **ignited.** This resulted in much greater power. The up-and-down movement of a piston in its cylinder is converted into circular motion by means of a **crankshaft.** To complete one revolution, or cycle, of the crankshaft, Otto's engine went through four stages, or **strokes,** of the piston. The engine is known as a four-stroke engine, and the process is called the **Otto cycle** after its inventor.

In a modern four-stroke engine, gasoline mixed with air is used as fuel. On the induction stroke the piston moves downwards drawing in

Gottlieb Daimler riding in his early automobile, 1886.

Traffic scene in Bangkok. There are over 200 million cars in the world today, and the number is growing. In most cities the streets are no wider than they were fifty years ago. So the traffic jams get worse and worse each year.

induction stroke compression stroke power stroke exhaust stroke

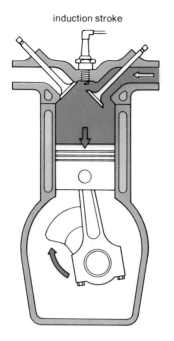
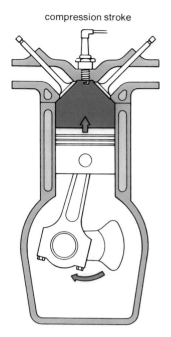
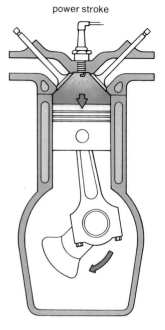
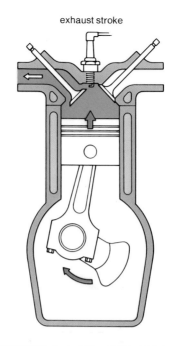

petrol vapor through the inlet valve. The valve closes and the piston moves up. As it does this, the gas is compressed in the top of the cylinder. With the valve still closed, the sparking plug ignites the gas causing a small explosion inside the cylinder. This provides the power stroke to push the piston down again, turning the crankshaft. As the piston starts to move up again the exhaust valve opens and the waste gases are forced out into the exhaust pipe.

One of the unpleasant things about the internal combustion or gas engine is that it is noisy. Also, waste gases are poured from the exhaust pipe into the environment. Apart from the noise, the exhaust gases include such poisons as lead compounds and carbon monoxide. These pollute the air and are damaging to health.

The Otto cycle is repeated several thousand times each minute when a car engine is running. In the lifetime of an average car a single piston in the engine may move up and down five hundred million times. With over two hundred million cars on the world's roads it is possible to see the size of the problem facing us.

There is no easy solution to the problem. Car manufacturers are beginning to introduce the energy-saving car. This has its own computer which tells the driver when he or she is using too much fuel. Some models even have computers to adjust the timing of the engine while it is running. In time alternative fuels which will not run out, such as hydrogen and methane, may have to be used instead of gasoline.

However, most experts believe that the best hope for the future lies in the development of the electrical car as an alternative to the gasoline

This test vehicle has an electric motor powered by batteries weighing 500 kg – considerably more than a tankful of gasoline. No less than 220 kg of lead-acid batteries are needed to equal the energy in one liter of gasoline. Weight and the limited range of batteries at present available are the main obstacles to successful development of the electric car.

There are several types of energy-saving driving aids to be found in modern cars. This design by Ford includes an advanced trip computer which displays fuel consumption. A map reading aid also shows route directions at each intersection.

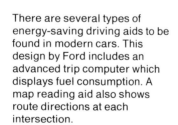

car. The chief problem which no one has yet succeeded in solving is the weight and the number of batteries needed to give enough power for other than short, low-speed runs. We may yet see electric cars in our cities. When we do, there will be less noise and less pollution of the environment.

First encounters with electricity

WHAT IS ELECTRICITY? No one invented it but the ancient Greeks knew something about it. Yet it is only in the last hundred years that we have learned how to put it to real use in the home and factory.

Electricity is a form of energy, but we cannot actually see it, hear it or smell it. We risk the danger of an electric shock if we try to touch it. We know about electricity by what it can do. The electric light bulb gives us light, the radio gives us sound and the electric stove cooks our food. These are some of the effects of electricity, but they do not tell us what it is.

There are two forms of electricity. One is called **static electricity**, which means that it stays in one place. The other is called **current electricity** because it flows.

We now have to go forward, a long way in time, to the sixteenth century. Sir William Gilbert was the doctor of Queen Elizabeth the First of England. In his spare time he did scientific experiments, and he found that other substances such as sulfur and glass could attract tiny objects just like amber. He called these substances 'electrics'. He believed that when these 'electric' substances were rubbed they became **charged** or filled with electricity. Sir William Gilbert was the man who first gave electricity its name.

The word electricity comes from the Greek word *elektron*, which means amber. Amber is hard, fossilized tree resin. It is known that more than 2000 years ago there lived a Greek philosopher called Thales. He noticed that when he rubbed an amber bead with a piece of silk cloth something strange happened. When the amber bead was placed near some small pieces of straw they jumped up to cling to the amber. He had discovered a new force. The straw was **attracted** to the amber.

In 1733 a French scientist, Charles du Fay, hung two amber beads on separate cotton threads. When the beads had been charged by rubbing each one with a cloth he placed them side by side. He was surprised to see that they **repelled** or pushed each other apart. Two glass beads when charged also repelled each other. But he was astonished to find that when a charged glass bead was put near a charged amber bead they attracted each other. Obviously the electric charge in the glass bead was some-

The ancient Greek scientist Thales (about 600 BC) noticed that a piece of amber, when rubbed with silk, would attract small pieces of straw. He had discovered static electricity.

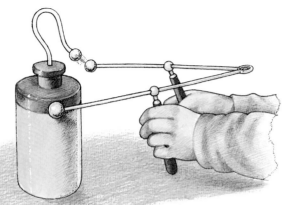

The Leyden jar was invented in the town of Leiden, Holland, in the middle of the eighteenth century. Electricity could be stored up when a charge of static electricity passed through a brass rod or wire into the jar containing water. The Leyden jar was the first capacitor.

how different to the charge in the amber bead.

By the middle of the eighteenth century the Leyden jar had been invented in the Dutch town of Leiden. This jar could actually store electricity and release it again. The device is today called a **capacitor.** It should not be confused with a battery which produces electricity by the use of chemicals. The inventor of the Leyden jar, Pieter van Musschenbroek, also discovered the dangerous effect of an electric shock. An accidental shock knocked him to the ground and terrified him.

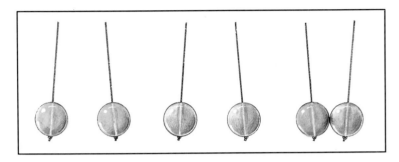

If two amber beads are suspended on string and then rubbed, they repel each other and are driven apart. The same applies with two glass beads. On the other hand if one amber bead and one glass bead are treated the same way, they attract each other. There are two kinds of electric charge, positive and negative.

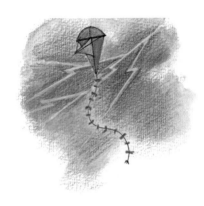

Benjamin Franklin, 1706–90. In 1752 he launched a kite in a thunderstorm. Electricity flowed down the wet line and created a spark on a metal key fastened near his hand. This led to the invention of the lightning conductor, which we see as a copper strip on the side of tall buildings. Franklin survived this dangerous experiment. Others who later tried the same thing were killed.

About the same time a famous American, Benjamin Franklin, and his son were risking their lives by flying a kite during a thunderstorm. Franklin had a metal door key tied near the lower end of the kite string and was able to see sparks jump between the key and his finger. This was a very dangerous experiment and some books tell of a Russian scientist who tried to do the same thing and was killed. The experiment led to the invention of the lightning conductor which is used for protecting tall buildings from lightning strikes.

Modern synthetic fibres and plastics easily become charged with static. You have probably noticed the crackles and sparks when you take off a nylon shirt or blouse in a darkened room. The spark as the charge leaks away is really a flow of current. Static and current electricity seem to have something in common.

The discovery of current electricity

IN 1791 an Italian by the name of Luigi Galvani, who lived and worked in Bologna, Italy, made a strange discovery. He was a **physiologist** (he taught the science of living things) and used frogs in his work. He noticed that touching a dead frog's leg with two different metals would make the muscles twitch.

Another Italian scientist, Alessandro Volta, carried the work further. He tested different metals moistened with salt. He finally produced a design of copper and zinc discs separated by pieces of cloth soaked in salt solution. This was called Volta's pile and when connected up, a small electric current was produced; he had invented the electric battery. Volta became quite famous and was summoned to Paris in 1801 to show his discoveries to Napoleon. It was a great step forward to produce an electric current as and when required. The scientist Sir Humphry Davy is famous for his invention of the coal miner's safety lamp. He did much research work in chemistry and built improved versions of Volta's battery to help him with his work. In 1810 he was able to demonstrate the first practical electric **arc lamp** at the Royal Institution, London. The effect of electricity on certain liquids interested him greatly. This process is called **electrolysis** and is used in the chemical industry. If an electric current is passed through water it can be broken down into the gases hydrogen and oxygen.

This painting shows Alessandro Volta (1745–1827) presenting one of the first electric batteries to Napoleon in 1801. He called it an 'artificial electrical organ', but it is commonly referred to as 'Volta's pile'. The electrical term 'volt' is named after his achievements. His invention helped create the electronic age.

The voltaic cell. Two rods, one copper and one zinc, are suspended in acid. When the two rods are connected by a wire a current flows.

This engraving of 1791 *(below)* is one of many that record the experiments of Luigi Galvani (1737–98). Following his chance discovery he showed that there was a link between electricity and the activity of muscles.

Magnetism was certainly known of in Ancient China, and magnets in the form of **lodestones** dug from the ground were used as compasses for navigation. The link between magnetism and electricity was discovered in 1819 by a Danish scientist, Hans Oersted. He found that if a wire carrying an electric current was placed near a magnetic compass it made the needle swing. This was the same effect that you get if you put an ordinary magnet near a compass needle. He had discovered the magnetic effect.

At about this time a young man called Michael Faraday was working for a bookseller. He was very interested in science and heard Humphry Davy lecture at the Royal Institution. He applied for a job there and was lucky to be made assistant to Davy at the age of twenty-one. Faraday had read of Oersted's discovery that an electric current can produce a magnetic effect. He was sure in his own mind that it should be possible to do it in reverse and have magnetism produce electricity. He always kept notebooks of his work and as early as 1822 we find in one of his notebooks the words 'convert magnetism into electricity'.

Faraday's laboratory at the Royal Institution, London. Michael Faraday (1791–1867) was the professor of chemistry at the Royal Institution, London, for more than thirty years. He is particularly famous, however, for his experiments in physics, including the work on electro-magnetism. Scientists some-times refer to him as 'the Father of Electricity'.

Faraday's electrolysis apparatus, 1834. Two electrodes connect to the glass globe and test tube (on top). The globe and tube are filled with water. When a current is passed through the electrodes, it changes the water into the gases hydrogen and oxygen. They rise and collect at the top of the tube.

Edison's light bulb (right). The American inventor, Thomas Edison (1847–1931), introduced the electric light bulb to the USA in 1879. It used a bamboo fibre covered with carbon. The inside of the bulb was a vacuum to prevent the filament from burning in air. Modern bulbs are filled with gases such as argon, that improve the glow.

In 1831 Faraday succeeded in producing an electric current by rotating a copper disc between the arms of a huge magnet. This was the first **dynamo**. The generators in our power stations that produce the power we use whenever we switch on a light are all based on Faraday's discoveries.

It would be unpleasant to return to the days before electric light. Yet it is less than 150 years since huge crowds gathered to see a demonstration of electric arc lighting from batteries. Arc lamps give off a very strong white light when an electric current makes a continuous spark or 'arc' between two carbon rods. One September evening in 1881 the townspeople of Godalming, England, saw their streets lit up. The lamps were electric and the power came from a generator driven by a water wheel.

Arc lamps had disadvantages. They became very hot and gave off smoke. An electric current flowing in certain types of wire can make it red hot. This is called the heating effect and is used in heaters and electric stoves. If the wire is made very thin it can become white hot and give us lighting.

The problem was to find a **filament** wire which did not burn away. Sir Joseph Swan in the United Kingdom and Thomas Edison in the United States both solved the problem in a similar way. They used a carbon filament inside a glass bulb from which most of the air had been removed. Electric lighting had arrived.

The nature of electricity

To UNDERSTAND the nature of electricity we must first look at atoms themselves. Everything is made of atoms which are so small that we cannot see them. When a beam of sunlight shines into a darkened room you can see the glint of tiny dust particles floating about. Each one of these is more than 50,000 atoms across, and even with the very best microscope we can only see particles 2 000 atoms across. In spite of this scientists have found that atoms themselves are made up of even smaller parts.

One of the best insulators is glass. This picture shows glass insulators on a junction tower, part of an electricity power line.

The hydrogen atom is the lightest atom, and the most simple. It has one proton, which forms the nucleus, and one electron in 'orbit'.

The carbon atom has six protons and six neutrons, which form the nucleus, as well as six electrons.

The copper atom has twenty-nine protons and twenty-nine electrons in orbit around the nucleus.

Atoms are not all alike and vary in mass and size. An atom can be imagined as a hard but fuzzy ball. Its center is called the **nucleus,** which is made up of **protons** and **neutrons. Electrons** move around the nucleus rather like space satellites in orbit around the Earth. The electrons are tiny but equal in number to the protons in the nucleus.

Electrons and protons carry an electric charge, positive (+) in the case of protons and negative (−) in the case of electrons. Each charge is equal, so one proton (+) cancels out one electron (−). Because each atom normally has an equal number of protons and electrons the atom remains electrically **neutral.** The neutrons in the nucleus carry no charge.

The gas hydrogen has the simplest atom of all, since each atom has only one proton and one electron. This is why hydrogen is so light and for this reason it was once used for filling balloons and airships. On the other hand copper is heavy. It has a nucleus with twenty-nine protons and twenty-nine electrons in orbit around the nucleus.

Electricians describe materials either as **conductors** or as **insulators** but what does this mean? Good conductors have atoms with a spare electron and these free electrons can dance from one atom to another. Insulators on the other hand have atoms which keep their electrons strictly in orbit.

The metal copper is a good conductor of heat and electricity. Atoms in metals are always arranged in regular patterns, and when a piece of copper is part of an electrical circuit the free electrons immediately start to flow in one direction only. A flow of electrons like this is called current electricity.

Metals such as copper and silver are good conductors. Non-metals such as plastics, wood and glass are insulators, but there are exceptions to this rule. The 'lead' from a pencil, which is not really lead, but graphite, will conduct electricity. Graphite is a type of soft carbon and it is a non-metal. There are also materials called **semiconductors** which are used for making **transistors.**

The nerves in our bodies operate by electricity, but instead of metal wires we have nerve fibres. The messages travel along the nerves at speeds between 1 and 100 meters per second, and our bodies operate at about 0.12 volt (volts are explained on the next page).

In a circuit using a battery the electrons flow from the battery around the circuit and return to the battery. Although a battery is not a pump, it seems to behave rather like one, and forces the electrons around the circuit. The moment the circuit is broken, by a switch or a break in the wire, the flow of electrons stops. When the battery has used up its store of chemical energy it stops 'pumping' and we say that the battery is flat.

We have learned that current electricity is a flow of electrons, but how can static be explained? Despite the fact that materials such as nylon and rubber are insulators, a charge can, nevertheless, build up on them. The clue lies in the old name for static. It used to be called **friction** electricity. This name was given to it

In a thunderstorm, lightning is caused by the movement of electrons from cloud to cloud (sheet lightning), or cloud to earth (fork lightning) as seen here.

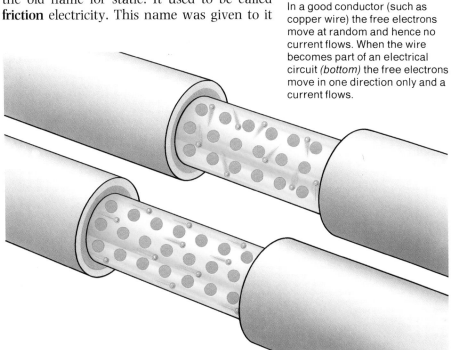

In a good conductor (such as copper wire) the free electrons move at random and hence no current flows. When the wire becomes part of an electrical circuit (bottom) the free electrons move in one direction only and a current flows.

because rubbing things together produces an electrical charge.

In a well-known experiment you can charge a plastic rod or comb by rubbing with a piece of cloth. Plastic is a good insulator, so all its atoms have a firm grip on their electrons. In the cloth, however, there are atoms with electrons not so strongly attached.

When the rod is rubbed with the cloth, some of the electrons are rubbed off the cloth and remain on the rod. As a result the rod builds up extra electrons, and therefore has a negative charge. The cloth has lost electrons so that it has extra protons. It has a positive charge.

Because they are so small we can only imagine electrons. This makes it all the more difficult to understand electricity. Yet it is the most important form of energy and we use it so much in our daily lives that we cannot easily do without it.

Volta, Ampère and Ohm

WE KNOW that an electric current is the flow of electrons along a conductor from one point to another. We also need to know the direction in which the current travels. Scientists in the past believed that electricity flowed from the positive (+) to the negative (−) terminal of a battery. When the electron was discovered, about one hundred years ago, scientists had to think again. It then became clear that the flow of electrons was in the opposite direction. They actually flow from the negative terminal of a battery, around the circuit, and back to the positive terminal. This is called **electron flow.**

We use electricity so much that we have to be able to measure it. Compare the flow of electrons along a wire with the flow of water and you will find it easier to understand. To make water flow through a pipe you need either the force of gravity or a pump to supply the pressure. In an electrical circuit the battery acts as an electron 'pump'. The electrical pressure is called **electromotive force** (EMF) and is measured in **volts** (named after Alessandro Volta).

The picture shows that if the amount of water in a container is doubled the pressure is doubled, and the water shoots out further. Compare this

Electric current is considered as flowing from the positive (+) to the negative (−). In fact the flow of electrons, called electron flow, is in the opposite direction, from negative to positive.

tremely dangerous. Voltage is measured with an instrument called a voltmeter.

In the conductor of a circuit, such as copper wire, the flow of electrons amongst the atoms is not completely smooth. There is **resistance** to the electron flow. Just as a large pipe lets water flow easily, so a thick wire allows electrons to do likewise. A thin pipe resists the flow of water, and a thin wire resists the flow of electrons. But we also need to know what the wire is made of. Copper wires have a low resistance and are used for the wiring in houses, shops and factories. Nichrome wire is a special high resistance wire used for the heating elements in stoves and heaters. It is an **alloy** and is made of two metals, nickel and chromium.

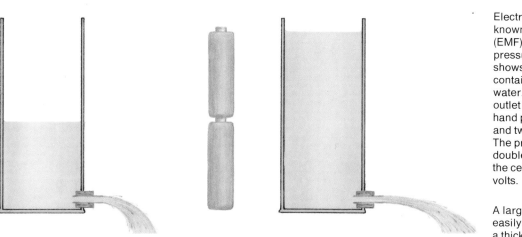

Electrical pressure in a circuit, known as electromotive force (EMF), can be compared to water pressure. The left-hand picture shows a single cell and a water container with a small amount of water. The pressure through the outlet pipe is weak. In the right-hand picture there are two cells and twice the amount of water. The pressure of the water is doubled. So also is the EMF in the cells. EMF is measured in volts.

A large pipe lets water flow more easily than a thin pipe. Likewise a thick conductor lets electricity flow more easily than a thin conductor.

with electricity. If we double up dry cells (a battery is only a battery if there are two or more cells) we double the voltage.

In experiments with electricity we should only use circuits working at low voltages, for example, four, six or twelve volts. Most household circuits are wired for 110 volts and are ex-

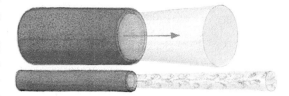

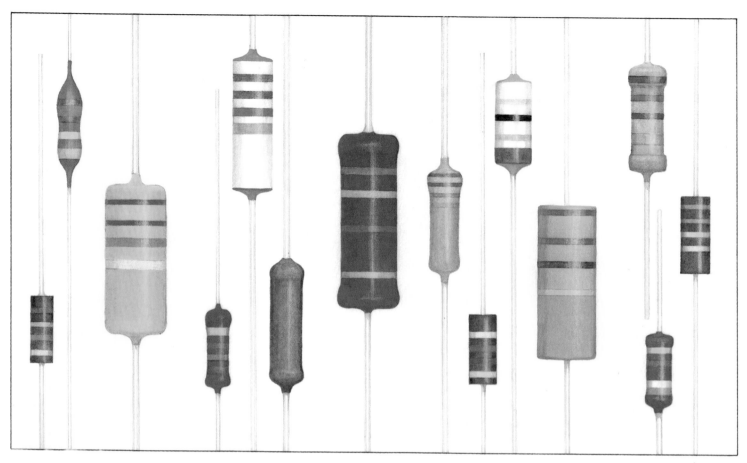

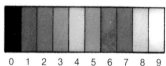

0 1 2 3 4 5 6 7 8 9

This illustration shows a variety of carbon resistors. There is an international color code (given above) so that the resistance value in ohms can be read off by anyone who needs to know. For example, red, blue, green and grey bands, in that order, represent a resistance of 2658 ohms. Below is shown an ordinary wire resistor.

The unit of resistance is called the **ohm** and is named after the German scientist Georg Simon Ohm. Many electrical products, such as radio and television sets, have parts known as resistors. A resistor may be made of carbon rod to an exact value of resistance, for example 140 ohm. In the illustration you can see how various values can be color-coded. Some resistors have the value printed in figures with the symbol Ω for ohm.

Measuring the flow of current is the same as counting the electrons as they whizz by. This is an impossible task since about six thousand billion might pass by in each second. In 1881, in memory of the French scientist André Ampère, the unit of electrical current was named the **ampere**. Ampère had carried on from the discoveries of Oersted. One ampere or amp is equal to that vast number of electrons passing by in one second, and current flow is measured with an ammeter.

These then are the units we use to measure electricity:

Electrical pressure or EMF
        is measured in volts
Electrical resistance
        is measured in ohms
Electrical current
        is measured in amperes

In 1826 George Ohm discovered that there was a simple connection between these three units. It was so useful that it was called Ohm's Law, and all simple electrical calculations are based on it.

OHM'S LAW

$$\text{Electrical current} = \frac{\text{electrical pressure}}{\text{electrical resistance}}$$

$$\text{or amperes} = \frac{\text{volts}}{\text{ohms}}$$

If any two of the three values are known you can work out the third. Electricity is not easy to understand, since we cannot see it. We have to make up for this by knowing how to measure it.

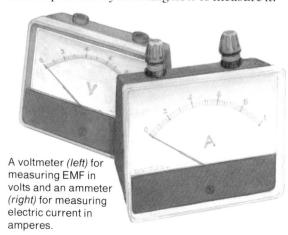

A voltmeter (left) for measuring EMF in volts and an ammeter (right) for measuring electric current in amperes.

Electricity and magnetism

IN THE YEAR 1600 Sir William Gilbert wrote a book about magnetism called *De Magnete*. He believed the Earth to be a giant magnet. To test this idea he made a sphere from lodestone, and placed small iron needles on its surface. To his amazement he found that all the needles pointed to the north and to the south.

The lodestone had been known to the Ancient Greeks and Chinese for its strange power of attracting things made of iron. It is a dark brown rock which is now called **magnetite.** Magnetite takes its name from Magnesia in Greece where it was found.

If you have a good magnet you can turn a

Sir William Gilbert, 1540–1603

A seventeenth-century lodestone, bound in brass. Lodestones were used on board ships to magnetize compass needles.

This diagram *(below)* shows how to turn a steel needle into a magnet.

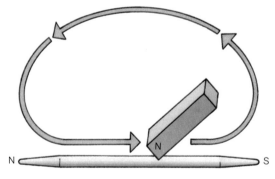

steel needle into a magnet by stroking it a few times as in the picture. The needle becomes magnetized. Inside a needle there are areas called **domains.** Each domain behaves as if it were made of even smaller magnets. Before the needle is magnetized the direction of each domain is different. After it has been stroked by the magnet, the domains in the needle all point the same way.

The **poles** of a magnet are often marked. They are the parts where the forces of attraction seem to be strongest. The pole which always tries to point north is called the north-seeking pole. The other is the south-seeking pole. Magnets are either attracted to each other or repelled, that is to say pushed away. There is a rule for this.

Unlike poles attract – like poles repel. The **magnetic field** of a magnet is the space around it where a magnetic force exists. This can be shown with iron filings as in the picture. The patterns are called **lines of flux.**

As Hans Christian Oersted had discovered in 1819, a wire through which an electric current is passing is always surrounded by a magnetic field. Later, an American scientist, Joseph Henry, wound copper wire into a coil and this produced a strong magnetic force as soon as an electric current was passed through it. Then William Sturgeon put an iron core inside a coil of wire and the magnetic force was even stronger. This kind of coil is called an electromagnet or **solenoid.** The magnetic field can be switched on

A magnetic field *(below)*. The lines of force that surround a magnet can be shown by the way iron filings arrange themselves.

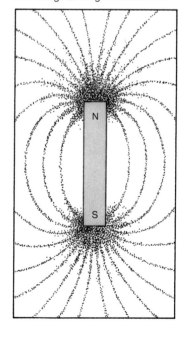

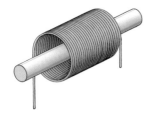

By passing electricity through a coil a magnetic force is produced. If an iron core is placed inside the coil a stronger magnetic force results.

or off as required. Electromagnets are widely used in telephones, electric bells and electric motors.

Oersted had shown how electricity could produce a magnetic field. It was Michael Faraday who showed how magnetism could produce electricity. On October 17, 1831, he produced an electric current by sliding a bar magnet into a long coil. Twelve days later he made a copper disk turn round between the poles of a very large magnet. He had made the first **dynamo** or **generator.**

Electric motors and generators are much the same in design. However, it is important to re-member that the one is the reverse of the other. Electric motors are powered by electricity. This makes the **rotor** revolve, providing mechanical power. When you switch on the washing machine, for example, the current makes the motor work. The motor drives the drum of the washing machine. Electrical energy is turned into mechanical energy.

Generators, on the other hand, are designed to make electricity. The rotor of the generator is made to revolve at a high speed by steam or mechanical power, and electricity is generated in the process. Mechanical energy is turned into electrical energy.

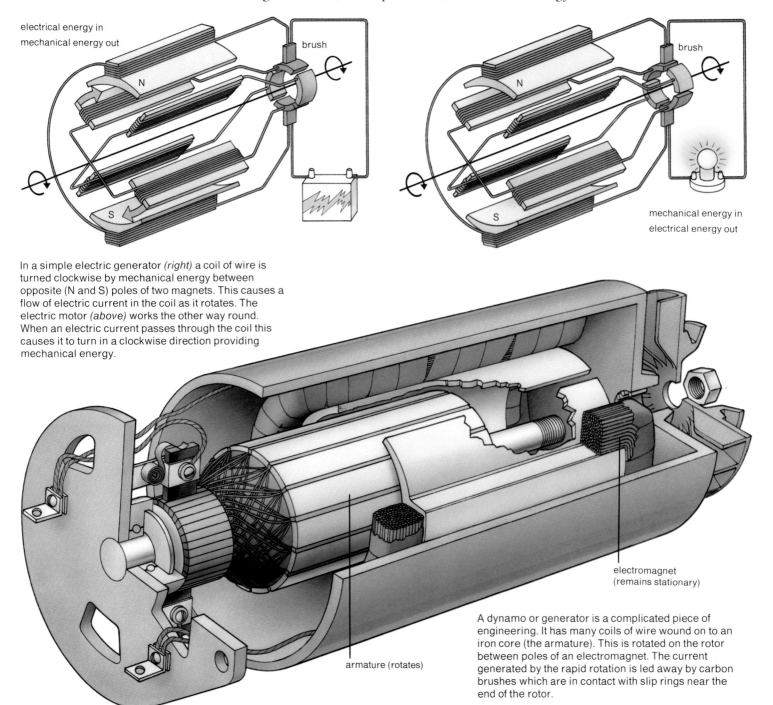

electrical energy in
mechanical energy out

brush

N

S

brush

N

S

mechanical energy in
electrical energy out

In a simple electric generator *(right)* a coil of wire is turned clockwise by mechanical energy between opposite (N and S) poles of two magnets. This causes a flow of electric current in the coil as it rotates. The electric motor *(above)* works the other way round. When an electric current passes through the coil this causes it to turn in a clockwise direction providing mechanical energy.

electromagnet
(remains stationary)

armature (rotates)

A dynamo or generator is a complicated piece of engineering. It has many coils of wire wound on to an iron core (the armature). This is rotated on the rotor between poles of an electromagnet. The current generated by the rapid rotation is led away by carbon brushes which are in contact with slip rings near the end of the rotor.

Power stations and electricity supply

ALTHOUGH MICHAEL FARADAY had discovered how to generate electricity in 1831, it took many years for the results of his work to be put to use. At first, electricity was used for lighting large houses and factories, each with its own generator. These machines all generated **direct current**, where the electric current flows in one direction only.

Then, in 1882, the first steam-driven power stations were built. In New York, Thomas A. Edison built the first large power station in the world. They were also built in London and on the South Coast of England. At first there were great problems. The authorities would not allow cables from the power stations to run either overhead or under the ground. For a while, cables ran along the gutters of the streets but this was very dangerous. Also, there were arguments whether power stations should supply direct current or **alternating current.**

The dynamo room at the first electric power station in New York, installed by Thomas Edison.

In 1889 the first large scale power station was built at Deptford in London. It was built by Ferranti who decided to use alternating current. Today all the public supply of electricity is by alternating current. This continually reverses, or alternates, its direction. The electricity generator which uses this system is known as an alternator.

In power stations the alternators are driven by steam turbines. The steam is produced by boilers heated by burning coal or oil, or from the heat of nuclear fission.

The first generators were powered by steam engines, which created a large amount of noise and vibration. As larger alternators were re-

quired the need arose for a smoother and more powerful engine. In 1884 an Englishman, Sir Charles Parsons, began to design a new kind of steam engine. He wanted an engine that would convert the energy of steam directly into rotating movement.

A windmill turns the energy of the wind directly into movement. Like a windmill, the steam turbine which Parsons designed had blades. Instead of four blades made of wood, his turbine had hundreds made of steel. First there was a circular row of fixed blades to guide the steam on to a set of revolving blades. Then the steam passed through a second row of guide blades on to another set of revolving blades.

This turbo-generator produces 660 megawatts (million watts) of electricity. It is seen here with its covers off, undergoing tests. Note the rows of blades.

These moving blades were fixed to the same shaft as the first. In all, the Parsons turbine had twenty-five rows of blades. All modern turbines work in much the same way.

Hydroelectric power stations use the energy of flowing water. The water is stored in a reservoir behind a dam. It is fed at great pressure down a pipe and on to a water turbine which spins the alternator. Hydroelectric power stations are expensive to build, but are cheap to run. They are used in those parts of the world where there is a good supply of flowing water.

Unlike the early power stations all modern ones are linked together by an overhead system of pylons and cables called the **grid.** This allows electricity to be sent over great distances and even between countries. We can only do this because we use alternating current, which has one great advantage. The voltage can easily be made greater or smaller by means of a **transformer.**

A transformer is made of two separate coils of wire wound on to a soft iron core. Depending on the number of turns on the primary (input) and the secondary (output) coils the voltage can be raised or lowered. Thus, if there are two hundred turns on the primary, and one hundred turns on the secondary, the voltage will be halved. A transformer will only work using alternating current.

Parson's steam turbine, designed in 1884, has proved the basis of the designs of turbines ever since. A row of revolving blades, not one wheel, create as much power as possible from the force of the steam.

The transformer is a vital part to any grid system. It is much cheaper and less wasteful of energy to carry high voltage electricity over long distances. Transformers at the power station are able to step up the voltage from 25,000 volts to 132,000 or more. Where the electricity is needed, other transformers lower the voltage to 110 volts or whatever voltage may be used in the locality.

Pylons, carrying high voltage cables are a common feature in the landscape.

Three stand-by gas turbines at a coal-fired power station. The great value of the gas turbine is that it can be brought up to full power in a matter of seconds.

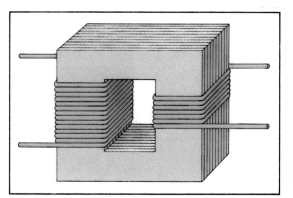

A simple transformer *(left)*. The core is made of soft iron and around it are wound two coils. Here the voltage in the left-hand wire will be double that of the right-hand wire, because there are twice as many turns. Voltage can be either stepped up or down as required.

A transformer sub-station *(right)*. This one converts 132,000 to 275,000 volts and is part of a national grid system in Great Britain.

Coal and coal gas

COAL IS MADE UP mainly of the decayed remains of the giant ferns and trees which covered parts of the Earth hundreds of millions of years ago. As the plants died they fell to the ground and rotted away to form a matted layer of soft material called **peat.** This can be found on the surface in wet moorland areas. Peat can be cut, dried and used as fuel. It is the first stage in the forming of coal.

As mud and sand built up over the peat, and millions of years went by, it was pressed down by the huge weight from above. It was turned into **lignite**, a soft brown coal. Further pressure from above and the heat beneath the Earth completed the change to produce layers of coal deep down which we call coal measures.

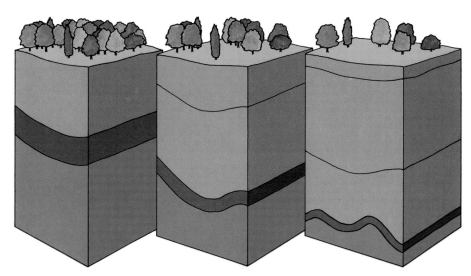

The formation of the coal that is mined today started in the Carboniferous Period, about 300 million years ago. Forests like this contained swampy vegetation which rotted down and in time formed peat. If successive layers of material such as sand or clay settled on the peat, the weight from above would eventually compress it into lignite. As the layers increased, the added weight, and the action of heat below the surface, eventually resulted in coal and, deeper still, in anthracite.

Coal varies in quality according to the amount of pressure and heat it has received during its formation. The lowest grade, **bituminous** or soft coal, burns easily and was once widely used for heating homes, although it produces a lot of smoke and soot. It is still used on open fires, but it is not allowed in many towns and cities. It causes **air pollution** which can be harmful to health.

The highest grade, **anthracite** or hard coal, does not burn well except at high temperatures. This makes it unsuitable for open fires. It is used in industrial furnaces and boilers, and in closed fires at home. For example it is used in solid fuel central heating systems. A draught of air is fed through it, supplying the oxygen that makes burning possible. Anthracite burns cleanly with little smoke or ash.

All the different types of coal are known as **fossil fuels.** Fossils are the remains of things which were once living. In a black lump of coal you can sometimes see the remains of a leaf which was green and alive long before men and women first walked the Earth.

Coal is graded according to how much carbon it contains. We call this its **carbon content.** The original plants were made up of about half carbon, and the rest of other chemicals including nitrogen, hydrogen and oxygen. During the change to peat, then to lignite, then to true coal, the other chemicals were gradually removed so that the carbon content increased. Carbon amounts to about half the contents of peat, more still of lignite, and about three-quarters that of soft coal. Hard coal has a high carbon content, up to 95 per cent in some cases.

A fossil of a fern *(left)* found in a coal measure in Yorkshire, England.

Modern coal-cutting shearer in action *(right)*. The water spray suppresses dust.

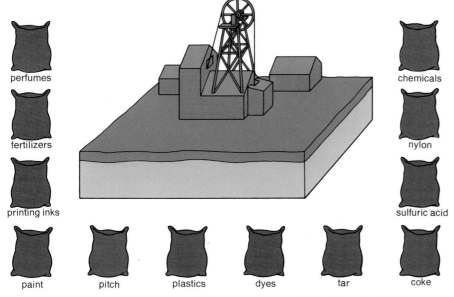

perfumes

fertilizers

printing inks

paint   pitch   plastics   dyes   tar   coke

chemicals

nylon

sulfuric acid

This diagram shows some of the by-products that are made from coal.

When coal is heated without air it does not burn but produces many useful **by-products.** One of these is coke, a fuel in its own right which can be used in homes and factories. Another by-product of coal is coal tar, which contains more than two hundred important raw materials. Tar can be used by industry to make soap, fats, dyes, plastics and perfumes.

The heating or 'cooking' process also releases coal gas which is made up of about half hydrogen, a quarter methane and a tenth nitrogen. Oxygen, carbon dioxide and other gases are present in smaller amounts.

Scientists living more than three hundred years ago knew that coal, when heated, would produce an **inflammable** gas. The idea that a gas could be lit and used for lighting was an exciting one and many experiments were carried out. Jean Tardier, a French chemist, found out how to take gas from coal as early as 1618. However, he could not use the gas for lighting, as he had intended, because the lamp he had designed was too dangerous.

By the turn of the nineteenth century gas was being made on a large scale in many countries throughout the world. It was made in towns and cities, and was stored in gasholders. The gas was then piped underground to homes and factories, for heating, lighting and power.

Coal gas produced this way is often called **town gas**, since it was made and used on the spot. It has now been largely replaced by **natural gas**, which is piped and transported direct from gas fields. Even so, the process of 'cooking' coal will continue since it releases valuable by-products and fuels, and leaves behind the solid fuel coke.

Coke is needed for the world's steel industry since it fulfils a number of purposes. When it has been heated white-hot it melts the iron, taking away the impurities in the process. It is also the chief source of carbon in steel.

It is quite wrong to suppose that the coal industry is declining. World coal production has doubled in the last thirty years and it is the most widespread of the fossil fuels. The coal mining industry is becoming more efficient, and safer for those who work the coal faces beneath the ground. There is enough coal in the world to last for hundreds of years.

Looking for oil

THE WORLD needs oil for cars, aircraft and ships. We burn it as fuel in power stations to make electricity. From oil we make such things as paint, plastics, clothing and animal food. More and more oil is needed as the world's population increases. It is hard to find and costly to produce. One day, perhaps in the not-too-distant future, little or no oil will be left. We have to use it carefully to make it last.

The search for oil goes on all the time, in many countries. It lies deep down under the ground or beneath the sea-bed. **Geologists**, whose job it is to look or prospect for oil, know that it is to be found in certain types of rock. It may lie under a desert, a jungle, or farmland. First, they look at air photographs. The shape of

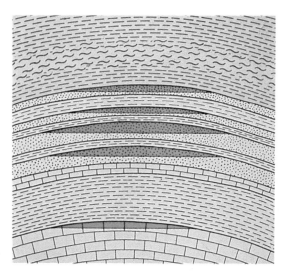

An anticline. The oil is trapped below the surface in strata of porous rock. Above, is a layer of very hard rock which does not allow the oil to pass through. Because of the curvature of the anticline, the oil and gas cannot escape horizontally.

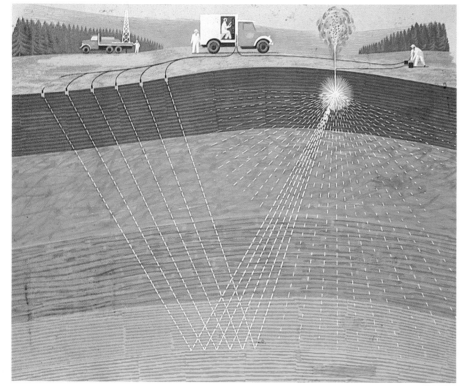

the land as seen from the air tells them about rock formation. Next, they go to a likely place and collect samples of rock to study. These samples tell them whether the rock beneath the surface is likely to hold oil, but they cannot be sure that they will find it.

When the oil is offshore and under the sea bed, exploration is carried out from a survey ship. Explosions send sound waves deep into the sea-bed. These echo back to an instrument in

When searching for oil, geologists use a seismograph. A small explosion is set off beneath the surface. Shock waves bounce back from the strata. These are recorded by the seismograph, which tells them how the layers are made up.

the ship and are recorded on a chart. The patterns made on the chart show whether the rocks are hard or soft, and where they lie. It is important to know this.

Oil, like coal, is a fossil fuel. It comes from the remains of plants and animals that lived in the shallow seas millions of years ago. As the living matter died it sank to the bottom and rotted. Layer upon layer of mud and sand piled on top and crushed the remains which gradually turned into oil. The oil was squeezed outwards until it could go no further. Oilfields are the burial grounds of living things from the dawn of life on Earth.

Oil may be trapped deep down in several ways. It can seep through soft rock like sandstone and limestone, but not through hard rock. When it meets hard rock, it is trapped. This may happen where the rock forms a huge arch, or **anticline**. The oil creeps up under the arch and cannot escape. A break, or **fault**, in a layer of rock may make a barrier which the oil cannot pass. Other rock shapes make underground pockets which trap the oil.

The oil explorers have to find these pockets of trapped oil. First a **derrick** is built. Then a test well has to be drilled deep into the earth. Derricks must be strong enough to hold the weight of a drill as it goes deeper and deeper through the rock.

The head of the drill, or **bit**, is fixed to a 10-m length of pipe. A powerful engine spins it into the earth at about four turns every second. New 10-m pipes are added as the bit goes deeper.

Every twenty or thirty hours the drill bit wears out and has to be changed. This means lifting the whole **drillstring**, as the chain of pipes is called, out of the ground. When the well is deep it may take ten or fifteen hours to change the bit, and in hard rock, therefore, drilling is a very slow process.

If the sides of the hole collapse the drillstring will be trapped. While drilling goes on, steel panels to protect the drillstring are lowered into place. This makes a strong well in which the drill can turn smoothly.

The drillstring is hollow and a special kind of mud is pumped down to keep the bit cool and

Test boring is another stage in the search for oil. Geologists examine the rock samples brought up by the rig. They then decide whether deeper, and more costly, drilling by an exploration rig is a worthwhile gamble.

help it cut. This mud flows back up the well, carrying bits of rock which are looked at for signs of oil. When oil is struck it tries to rush to the surface. In a **blowout** the oil escapes with such force that it can kill men and damage the drilling rig. Every well has a strong cap called a blowout preventer to reduce the risk of oil gushing out.

If oil is struck, the oil company must decide whether to develop the oilfield. They will have to find out how much oil there is and what it will cost to get it out. Also, it is important to know whether the oil will flow easily and whether it is of good quality.

A drill bit is examined on board a production platform in the North Sea. Despite being made of the hardest material, including diamond, bits can wear out very quickly if drilling through hard rock.

The drilling of wells is an activity calling for planning and team-work. When the drill bit is changed, the drill pipe has to be raised and unscrewed with the use of power tongs (right).

45

Offshore rigs

NEARLY HALF OUR OIL is found under the sea. Obtaining this oil is made possible by offshore rigs. These are among the largest structures ever built. They are platforms high above the sea.

There are two kinds of offshore rig. The first is the drilling rig, which is used to find the oil. The second kind, the production platform, is a gathering station where the oil from several wells is collected.

The simplest drilling rigs stand on legs on the sea-bed and are fixed in place. Another type stands on the sea-bed but can be moved. The largest are floating platforms held by chains to the sea-bed.

There are several hundred drilling rigs around the world. Most are working above the **continental shelf**, the shallow part of the sea close to land. Beyond the continental shelf, the sea-bed falls steeply and is too deep for exploration at present. The main areas of offshore exploration are the southern coast of the United States, the Middle East, the Far East, and the North Sea. Many rich oilfields have been found in the North Sea, between Britain and the mainland of Europe.

Oil companies do not usually own drilling rigs. They hire one when they want to sink a test well. A typical drilling rig stands about

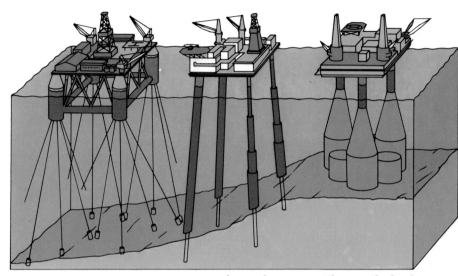

Various types of offshore rigs (not to scale). Semi-submersible rig *(left)*, floats on the surface and is firmly anchored to the sea bed. The Magnus *(center)* is fixed by huge steel legs. The production platform *(right)*, collects oil from several wells and stores it in massive tanks at the base.

25 m above the water. The top deck of a rig is often bigger than a football field, and has its own helicopter pad and lifeboats. Seventy or eighty men may live on the rig for several weeks at a time. They include the drilling team, divers, cooks, and others who look after the day-to-day running of the rig.

The divers repair and set up equipment on the sea-bed, where the drill bites the earth. Another part of their job is to inspect the underwater parts of the rig for wear and damage.

The search for oil never stops and the rigs must work all year round. This means that a rig must be strong enough to face winter storms. It has to remain steady in the water, even when the waves are 25 m high.

Oilmen can drill in water over 1000 m deep. Wells do not have to be drilled straight down. By using a tool called a **whipstock**, oilmen can bend their drillstrings outwards. In this way, they can reach all round an oilfield. Several wells can be linked to one production platform.

When oil is ready to flow, it must be kept under control. A group of pipes and valves called a 'Christmas tree' is fixed to the well head. This collects the oil into a pipeline which takes it to the production platform. Here, water and gas are removed from the oil and it is ready to be sent ashore. It may be piped to an oil terminal on the shore, or to an oil storage buoy near the

Cormorant 'A' production platform being towed out of Stord Fjord, Norway, at the beginning of a 250-mile journey to its North Sea location.

Dusk view of production platform with choppy sea in the foreground. This concrete structure operates in a water depth of 139 m.

A drilling rig. This type is called a semi-submersible. It is anchored in position by chains and has large air tanks below the surface that are designed to keep the rig as stable as possible. Although rigs like this one are vast in size they have to pack in much more than just the oil-drilling equipment itself. Seventy or more men work on the rig, including cooks and deep-sea divers. There are cranes to unload boats bringing supplies, and a pad for helicopter landings. Lifeboats are there in case of emergency.

rig. Tankers load oil at the buoy and carry it ashore. North Sea oil is often carried ashore by tankers.

Production platforms are like man-made islands. They stand on the sea-bed and rise high above the water. Ninian Central platform, in the North Sea, is nearly as tall as the Eiffel Tower – about 230 m. It was built on land and towed by tugs to the oilfield. Production platforms are built in deep-water harbors, as in Scotland and Norway, before being towed out to sea.

Oilmen on offshore rigs cannot travel to work every day. They must live on the rigs. Work goes on all the time, in shifts. When one group of men finishes, another group takes over straight away. Shifts are usually twelve hours long. A lot of the work is heavy, hard and dirty, but the men can relax. The oil companies know that the men may get bored and must be well looked after. Living and sleeping rooms are comfortable and the food is good. Fresh food is brought in regularly by helicopter and supply ships. There are games rooms, libraries, television and video, but the men miss their families and friends. After several weeks on the rig, they spend two or three weeks at home. They are paid well and have plenty of time off, which helps to make up for their tough and sometimes dangerous work.

Oil refining and by-products

THE CRUDE OIL that flows from a well is black and thick. Before it can be used it must be taken to a **refinery**, where it is cleaned and changed into different types of oil product. Crude oil was once refined close to the oilfields. Now it is usually taken direct to refineries in the countries which will use it.

Huge oil tankers carry the crude oil across the seas of the world. The oil-carrying space in these ships is divided into separate tanks for safety reasons and so that different kinds of oil can be carried. Also, it prevents oil from surging about in high seas. The engines and the crew's living quarters are at the rear or after end of the ship, well away from the oil. This reduces the risk of fire.

Supertankers, which can carry 500,000 metric tons of oil, are so big that the crew often use bicycles to get from one end to the other. Loading and unloading takes place at special deepwater terminals. Oil fumes inside the tanks can easily catch fire and the men have to take very great care when doing this work.

America's largest oil terminal is the Offshore Oil Port in Louisiana. Crude oil arrives by supertanker from the oilfields and is transferred to the mainland by pipeline. It is then distributed to the refineries.

The Sullom Voe oil terminal, in the Shetland Isles, Scotland. Oil arrives by pipeline and tankers from North Sea fields. From here the supertankers deliver the oil to Europe and the USA.

The best way of moving oil over land is by pipeline. Pumping stations keep the oil flowing at between 5 and 7 km per hour and the pipelines cover great distances. The longest pipeline, 4300 km, runs from Alberta in Canada to Buffalo, New York. Countries using a lot of oil are criss-crossed by pipelines, most of them underground. More than half the length of the Trans-Alaska pipeline (1300 km) is above the ground on stilts.

Pipelines are usually about 1 m wide. They are made of lengths of steel pipe about 12 m long, welded together and sealed against leaks. The pipe is protected from rust, then lowered into a trench. The soil is replaced and crops can grow again. Oil companies are proud of the way in which their pipelines are built without harm to the environment.

When the crude oil reaches the refinery, it is pumped into storage tanks. It is a mixture of many liquids, which have to be separated from each other and cleaned. First, it is heated to

A night view of an oil refinery at Lima, Peru. Since modern refineries employ the latest technology, few people are needed to run this dramatic looking complex.

400°C. As it boils, it gives off gases which pass into tall towers called **distillation** columns. The gases cool as they rise inside the columns and change back to liquids. This is called **condensation.**

The columns are sometimes known as **fractionating** columns since they separate the crude oil into different parts, or fractions. The lightest fractions are at the top and the heaviest at the bottom of the column. The fractions are caught in trays at various levels and piped away to storage tanks. Light fractions are needed most, but first distillation produces a number of heavy fractions. Scientists have found a way of changing heavy to light fractions. They use a chemical called a **catalyst** to change, or crack, the heavy fractions. The process is known as catalytic cracking.

A range of by-products leaves the refinery. Gas from the top of the columns is bottled for use in homes, and for trailer owners and campers. Next comes gasoline, the product we know best. It is taken from the refinery by road, rail and pipeline. Kerosine is produced below gasoline in the column and is the fuel for jet aircraft. Lower down the column comes diesel fuel for trains, trucks, buses and some cars. Oil for home heating also comes from the same level. Lubricating oil, the oil which makes machines move easily, comes next, with wax for candles and polishes.

The heavy fractions at the base of the column give oil for use as fuel for ships and factories. It is thicker than diesel oil. Heaviest of all is bitumen, a black sticky substance used for roadmaking, roofing materials, and for waterproofing.

Some of the fractions mix easily with each other and with other chemicals. By mixing them, it is possible to make plastic materials which are used in hundreds of different ways. Plastics such as **polyurethane** are used to make paint, and builders use **polypropylene** pipes for gutters and drains. Some of the clothes you are wearing are probably made of plastic fibres.

Chemicals made from oil are in daily use in washing powders and liquids. Farmers and gardeners use fertilizers and weedkillers which may be oil-based products. The list of oil products and chemicals is almost unending. They are known as **hydrocarbons**, since the chemicals carbon and hydrogen are in all of them. All living matter contains carbon, and we are reminded of the fact that oil and its products started as plants and small animals millions of years ago.

A distillation column. The crude oil is heated and the fractions separate at various levels. The lightest gas rises to the top, and the heaviest oil stays at the bottom.

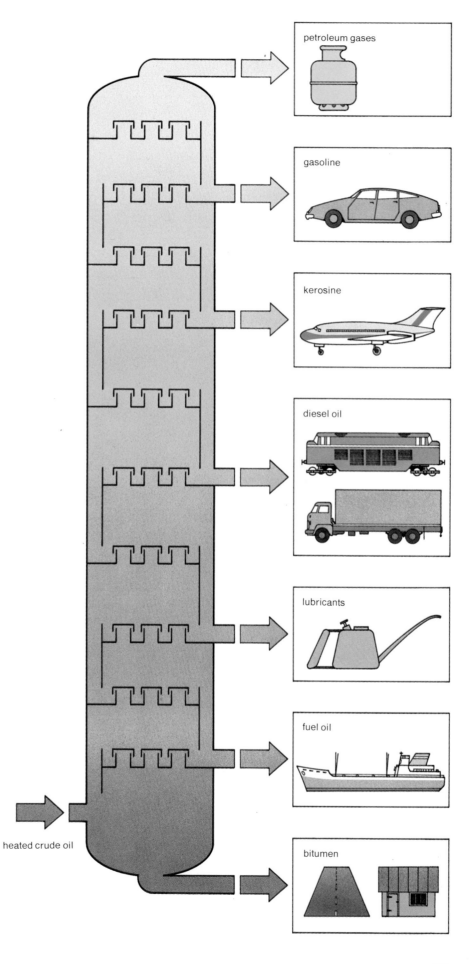

petroleum gases

gasoline

kerosine

diesel oil

lubricants

fuel oil

heated crude oil

bitumen

Nuclear power

STRANGE AS IT MAY SEEM, a nuclear power station uses steam turbines to drive the alternators just like any large coal-fired power station. It is only the source of heat that is different. At the heart of a nuclear power station is the **reactor.** This is a large vessel in which vast amounts of heat are produced by a process called **nuclear fission.** The heat caused by fission is used to make steam to drive the turbines.

To understand nuclear fission, or the splitting of atoms, we need to know more about atoms themselves. Everything is made of atoms. Atoms are so very small that we can never see them. There are billions of atoms in a single drop of water. For many years scientists believed that atoms were the smallest objects that could possibly exist. The word atom comes from a Greek word meaning something that cannot be divided or split apart.

However, in 1905, Albert Einstein put forward a remarkable idea. He stated that if an atom could in some way be changed or split, then an enormous amount of energy would be released. It was Ernest Rutherford who was the first scientist to succeed in splitting or changing an atom. This famous experiment took place at the Cavendish Laboratory in Cambridge, England, in 1919. Rutherford and other scientists then set out to prove that atoms are made of even smaller parts, or **particles.**

We now know that most of an atom is empty space. At the center of the atom is the nucleus.

Hydrogen, a gas, is the simplest element and it is the lightest gas. Its nucleus is a single proton. It is the only element without a neutron. A single electron orbits around the nucleus.

Ernest Rutherford 1871–1937, was a New Zealander by birth. It was in England that he did most of the work for which he is famous.

Albert Einstein 1879–1955. When he was only 26, Einstein published his special Theory of Relativity, connecting mass with energy.

Orbiting around the nucleus like satellites are one or more tiny electrons. The nucleus itself is made up of protons and neutrons. Hydrogen is a very light gas and its atoms are the simplest of all. Each hydrogen atom has one proton as its nucleus and one electron in orbit. In contrast **uranium** is a very heavy metal. Each uranium atom has ninety-two protons and more than one hundred neutrons in its nucleus. It also has ninety-two electrons in orbit.

Uranium is the only natural material in which fission will occur easily. When struck by a neutron the nucleus of an uranium atom splits into two roughly equal pieces. These pieces move apart at high speed. At the same time two or three free neutrons shoot off to collide with other uranium atoms. When they collide great heat is produced and more fission takes place. We call this a chain reaction. It can, as a result, release continuous energy.

As well as heat energy, **radioactivity** is created during fission. The rays are dangerous and penetrating. A very thick steel and concrete shield around the reactor stops these rays escaping. The uranium fuel is placed in long tubes which are packed together in the **core,** or

neutron   uranium
atom

Nuclear fission takes place when an uranium atom is struck by a neutron. The uranium nucleus divides into two roughly equal parts forming atoms of different elements. Two or three free neutrons are released. As these neutrons collide with other uranium atoms the process is repeated and multiplied. A chain reaction takes place and vast energy is released.

Oldbury Nuclear Power Station in England was one of the first nuclear stations to use reactors enclosed in concrete pressure vessels.

In an Advanced Gas-cooled Reactor, fuel rods are arranged in the core so that gas can flow past to remove the heat. This boils water, producing high pressure steam which is used to drive a turbine in the generator. The nuclear reaction is controlled by control rods of a material such as boron. This absorbs neutrons and slows down the chain reaction. The reactor is surrounded by a shield of concrete or steel.

Interior of a reactor hall at a nuclear power station.

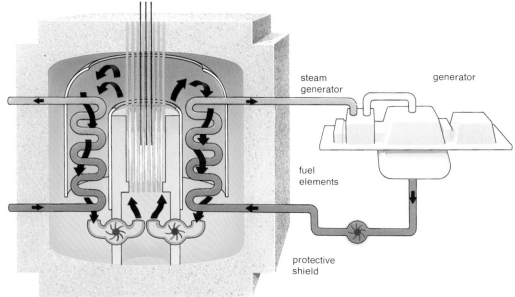

steam generator

generator

fuel elements

protective shield

center of the reactor. Just as we need to be able to turn the heat up or down on a stove, we need to control the nuclear fission. This is done by rods of boron or cadmium which can be lowered into the reactor. These rods soak up neutrons and slow down the chain reaction. If the reactor becomes too hot the rods are pushed right down and in this way they can stop the reaction altogether.

It is important to be able to remove the heat from inside the reactor. In order to do this a gas or a liquid is pumped through. This heat is then used to boil water in a **heat exchanger.** You may

have a small heat exchanger in your basement. This is where hot water from the boiler heats water for your bath. In a power station heat exchanger the water is boiled to make steam for the turbines. The turbines drive the alternators which make the electricity.

As with all power stations, waste material in the form of spent fuel is produced. Because this waste remains radioactive for many years it has to be treated with great care. Scientists are now working on a system to make the waste into solid glass blocks which can be stored until no longer radioactive.

Some scientists hope that in the future it will be possible to replace fission by **fusion.** In the fusion process light atoms such as hydrogen would join together releasing vast energy, and no waste would be produced. One day scientists may solve this problem. When they do the world will have an almost unending source of energy. Uranium is scarce and expensive to mine, but there will never be a shortage of hydrogen, since it is hydrogen and oxygen which make up water.

Pollution of the environment

ALL ANIMALS AND PLANTS produce waste materials in one form or another. Nature takes care of all these wastes by turning them back into the raw materials that can be used by plants. Humans are the first animals to produce waste in such large amounts, or of such a type, that nature cannot keep up with the task of breaking them down and making them harmless. We call these materials **pollutants**, whether or not they can eventually be broken down naturally.

The human race has always had a problem with waste and rubbish. Even cavemen had special areas set aside on which they dumped all their uneaten food, bones and shells. Many rubbish dumps left by the Romans are now a rich source of jewelry and pottery fragments. The difference between then and now is that the world population is much larger and produces a growing amount of waste.

In the modern world, pollution is split into types, depending on where the pollution occurs. Car exhausts and the fumes from chimneys cause **air pollution.** Chemical factories, the use of farm crop sprays and the disposal of sewage can lead to **water pollution.** We also have special names for certain types of pollution. The most common one is **oil pollution**, caused by accidents to oil tankers or drilling rigs at sea.

It is easy to forget something that we throw away. However, increasing amounts of waste from our everyday lives end up in dumps like this.

A lot of air pollution is caused by factories which emit dangerous smoke. The gradual build-up of poisonous fumes in the atmosphere can harm humans and wildlife.

The burning of coal as a household fuel is now banned in many cities and towns. The black, grimy outside of buildings is due to the soot, and to a gas called **sulfur dioxide**, in smoke. These same fumes caused many people to die of lung diseases before smokeless fuels were developed.

Many factories produce chemical wastes which are often poured into a river or put into containers and dumped at sea. In many parts of the world, untreated – or raw – sewage is pumped straight into the sea. In small amounts this is perfectly safe, but with the growth of our cities the increased amount of sewage is causing disease. In many rivers and seas, plant and animal life is dwindling because of pollution.

Even the countryside can be polluted. Farmers use **fertilizer** on their fields to improve the crop yield. One of the most common fertilizers is a chemical called **phosphate** which encourages

A polluted river may become a dead one. Waste discolors the water, may cause silting, and can kill off all life within it. Care can prevent this happening.

plant growth. As water from the fields drains into streams and rivers, the phosphate in the water causes rapid growth of green, slimy plants called **algae.** These tiny plants grow so quickly in the rich fertilizer 'soup' that the oxygen in the rivers is used up. This leads to the death of large numbers of water plants and animals because of a lack of oxygen.

Chemicals called **insecticides** are often sprayed on fields to kill insect pests. These sprays also kill useful insects such as honey-bees. One insecticide called DDT was once widely used to keep down mosquitoes. Once in the soil, DDT cannot be destroyed. It was banned because it was slowly poisoning the environment.

In the past fifteen years there have been at least two major oil-tanker accidents. These have released millions of metric tons of crude-oil on to the surface of the sea in huge pools called oil-slicks. The oil floats on the sea and is eventually washed ashore where it kills thousands of sea-birds and much of the seashore life.

An unusual sort of pollution is caused by aerosol cans, rockets and high-flying jet airplanes. The upper part of the atmosphere includes a thin layer of a gas called **ozone.** This is a type of oxygen which absorbs the harmful ultraviolet rays from space. It is Earth's protective shield. Without this layer all life would soon be destroyed by the harmful radiations. The gas which provides the pressure in aerosol cans reacts with ozone and breaks it down. High flying aircraft and rockets also burn up the ozone and further reduce the protective layer.

Oil pollution often makes the news. It can be caused deliberately by ships cleaning out their tanks, or by accident. Attempts at clearing it up are very costly and the destruction of sea-birds is often heavy.

With increasing technology, we are producing more and more harmful wastes. Perhaps the most dangerous are the radioactive wastes from nuclear reactors or from atomic weapon testing. These radioactive wastes need thousands of years before they are 'safe'. In the future we must be prepared to spend a lot of money and time in finding better ways of dealing with all our waste products before we pollute and damage our whole environment.

Water pollution easily occurs when waste chemicals are washed away or handled carelessly. Pollution in lakes, rivers and areas of the sea soon kills the fish.

Conservation

CONSERVATION is a word that means different things to different people. Wildlife or nature conservation is when we try to prevent animals and plants becoming extinct. To the factory owner the word conservation means the re-using, or **recycling,** of waste materials to produce new goods. In the home we practice energy conservation. This is when we try to use as little fuel or electricity as possible. Generally, conservation means keeping together, or preserving those parts of the natural world we can use. These are called our natural resources.

One of the greatest problems in the world today is overcrowding. Our planet seems very large but one day it will be unable to support any more people. The increasing number of people in our world is the reason why conservation is so important. Unless we conserve our fuel and our resources, we will soon run out of them. The continued destruction of huge areas of natural forests, grasslands and lakes could lead to the decline and even the extinction of the human race.

We are part of the natural world and we need its great variety for our own well being. Mankind has already chopped down thirty-five per cent of all the world's forests, and has been responsible for the extinction of at least thirty-six species of mammal and ninety-four species of bird.

The forests are a good example of how conservation is needed in the world today. The first cavemen took firewood from the forests as fuel for their fires. Even today firewood is still the main source of fuel for cooking in many parts of the world. In Europe we have already lost that resource as nearly all the natural forests have been chopped down to make farmland for growing more food. Today, the last great forests of the world, the **tropical rain forests,** are being cleared very quickly. These forests are being cut down for the same reasons the European forests were felled. The aim is to make new farmland and feed the ever increasing number of people. Unfortunately, the high rainfall and the poor soils of rain forest areas mean that the new farmland becomes useless after only a few years.

Clearing great areas of rain forest will also affect the climate of the area. These forests act like giant sponges, soaking up the very high rainfall of the tropics. Most of this water

The increasing demand for energy means large power stations have to be built. As fossil fuel supplies begin to run out, we might well return to using the same source of power as this old windmill. This reminds us of the past when energy demands were not so great.

The giant panda, itself a rare animal, is the symbol of the World Wildlife Fund, which was formed for the conservation of animals, and the protection of rare species.

The beginning of the Trans-Amazonian highway. Access to dense areas of tropical rain forest can result in widespread forest clearance. Unless great care is taken whole areas may become desert.

Strip mining destroys large areas of countryside. The scars that are left are sometimes used as giant waste dumps, or may be turned into man-made lakes.

evaporates back into the atmosphere to form yet more clouds. The whole weather pattern of these regions relies on the return of water from the forests. Once the forest is chopped down the rain-water will either run off the land into rivers or soak into the ground. This upset of the water balance can change fertile rain-forest areas into desert.

In our homes the conservation of fuel begins at the electric light switch. Each unit of electricity that flows into our homes is produced by a power station. In the power station fuels such as coal or oil are burned to release their energy. This energy is changed into electrical energy by turbines and can then be used by us in our homes. The natural resources of coal, oil and gas are limited. Fossil fuel resources cannot be replaced but careful use will make our stocks last much longer. By thinking whether or not you need an electric light on, or by switching it off when you leave a room, you are helping to conserve the fuel resources of the world.

Modern homes are built to conserve fuel by using building materials and methods that **insulate** the house. In the colder parts of the world insulation helps to keep heat in and save fuel. In hot climates the insulation helps to keep heat out and avoids using energy to cool the house. Different peoples have different ways of insulating their homes. Many western countries use extra sheets of glass in their windows; this is called **double-glazing.** Air is trapped between the sheets of glass and prevents either heat or cold from passing through.

Energy from Sun and wind

WE KNOW that huge amounts of energy come from the Sun. Over 2000 years ago the Romans were using crude forms of solar energy to heat their houses. In the nineteenth century the first solar-powered engines and refrigerators were invented. Thousands of homes in the USA had solar water heaters during the 1890s.

During the first half of this century solar energy was forgotten. First coal and then oil and gas, the fossil fuels, were so cheap in the industrial countries of the world that other sources of energy were ignored. However, some scientists could see that the fossil fuels would not last for ever. By 1950 new work had started on methods of getting heat, power and electricity from the Sun. Today, almost all countries use some solar energy, and many have plans to use larger amounts in the future.

The way we make most use of solar radiation is to convert it into heat. This is usually done in a **flat plate collector**, which is often a glass-

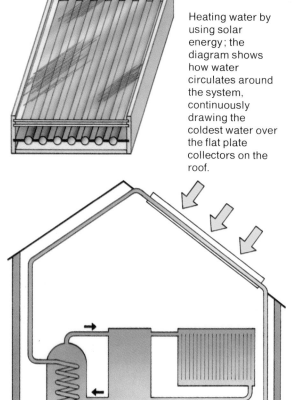

A solar panel

Heating water by using solar energy; the diagram shows how water circulates around the system, continuously drawing the coldest water over the flat plate collectors on the roof.

topped box with a black plate inside. The plate has tubes or passages through which water can pass. The water is heated during the day by the Sun's rays. It can be stored for use overnight in an insulated tank. For heating buildings it is possible to store the heat for longer periods in large heat-storage tanks. In Sweden, for example, groups of houses have been built that are connected to large solar-storage tanks. The houses are heated almost entirely by stored solar heat throughout the year.

High temperatures can be obtained by using

The solar furnace at Odeillo, France. Temperatures up to 4000°C can be reached by a complex system of mirrors that focus the sun's rays on to a furnace. Odeillo is not a power station, but it uses its great heat to melt metals.

mirrors to focus the rays of the Sun. The first large system was at Odeillo in France. Recently, some systems ten times larger have been built in the USA. The advantage of these high temperature plants is that they can be used to power engines, for water pumping and for running machines.

The rays of the Sun are also used for **distillation**, to produce pure water from salt or dirty water. The dirty water is heated in a glass-covered trough and the steam **condenses** on the glass cover to give pure water.

Solar energy can also be converted directly into electricity. The earliest use was in the first spacecraft, but **solar cells**, as they are called, are now being used for other purposes. These include power for radio and TV in remote districts, refrigerators, buoys and other navigation aids. Although they are costly today, they will become cheaper to make as new methods are found. Some engineers can predict the day when the roof of a new house could be both a water heater and an electricity generator – all from solar power.

Much of the Sun's energy goes to heat up the air close to the Earth's surface. This causes the winds, so the energy from wind also comes indirectly from the Sun. Wind has been used as a source of power for thousands of years, both on land and at sea. Early windmills were first used about five hundred years ago. The sails were supported on a tower and the windmill was mainly used for grinding corn. The other main use was for pumping water for crops in fields.

Wind power was first used for making electricity in 1890. Over the next fifty years thous-

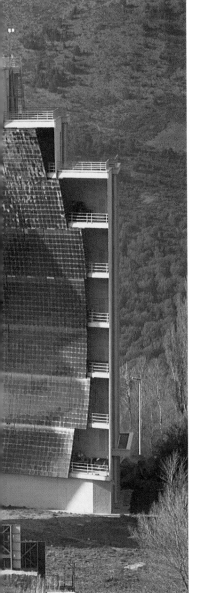

An example of the advantages of solar power. This buoy, off the coast of Australia, uses solar cells to power its transmitter. The radio beacon acts as an aid to navigation, and needs no refuelling.

Latest designs in wind turbines use blades rather than the old-fashioned sails of early windmills. The diameter of the blades is much larger.

ands of small windmills were built to make electricity. In 1941 the first windmill capable of producing over 1000 kilowatts – one kilowatt is a thousand watts – was built in the United States. Today, with better materials, there are a number of windmills in the world each producing up to 3000 kilowatts.

The power given by wind depends upon the wind speed. It is very important to find a site with a high average wind speed. If one site has an average wind speed twice that of another site, then a windmill at the first site will produce eight times as much energy compared to a similar windmill at the second site. Blade diameter is also important.

Compared with many other sources of energy, wind power has three main advantages. The first is that a modern windmill can be built quickly. A nuclear power station or a tidal barrage could take over ten years to build. The second is that, for many countries, the highest winds occur during the winter, when electricity is needed most. The third is that wind is a renewable source of energy, available throughout the world.

There are some disadvantages. They must stand up to the strongest winds – gusts can go up to 150 km per hour. Also they can be noisy. Some people think they are ugly, but are they any more ugly than modern power stations with lines of large pylons?

Energy from water and the Earth

THE ENERGY in streams of fast flowing water was used over 2000 years ago by the Romans. The first water mills were for grinding corn, but later they were used in factories and for pumping water from mines. The flowing water was taken through a channel to a large wheel fitted with flat blades. Each blade, in turn, was pushed round by the running water so that the main shaft could provide power. However, they were not very efficient at changing the energy in the water into power. The wheels were turning too slowly to make electricity.

An early water mill with vertical rollers used for crushing sugar cane, powered by an overshot water wheel.

A cut-away diagram of a hydroelectric turbine, showing how the flow of water spins the blades of the turbine as it passes.

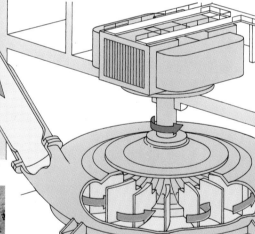

Modern water turbines use different methods. They have curved internal blades and the water flows at high speeds through pipes, rather than through open channels. **Hydroelectricity** is the name given to electricity made from the power of water. The world's first 'hydro' system was set up in 1882 in Wisconsin. It had a very small output, giving power for only 250 light bulbs. However, from this small start the production of hydroelectricity has grown steadily so that today it provides over six per cent of the world's energy needs.

There are some disadvantages. Large dams must be built to hold the water before it flows to the turbines. The environment would suffer if the flow of a river was completely stopped. Also, small pieces of earth and sand are trapped in the still water at a dam. As these deposits build up

Hoover Dam, on the Colorado River. Situated in a narrow, rocky valley, the dam is well supported by the valley walls. Dams are designed to withstand massive pressures which build up behind their walls.

problems are caused. For example, one reservoir in Colombia lost 80 per cent of its water capacity in only fifteen years for this reason. Not every country has suitable sites for development. The main advantages are that hydro power is long lasting and reliable. Maintenance is easy and the energy is renewable. Compared with fossil fuels, hydroelectric schemes are much less expensive.

Energy obtained from the ebb and flow of the tides could be very large. In practice, it is not easy to find good sites. Tidal mills were used for hundreds of years in some European countries. In the British Isles there were over one hundred working in the nineteenth century. They produced similar amounts of power to the early windmills – less than 50 kilowatts. They, also, were often used for grinding corn.

Tidal power station on the Rance estuary, France. Power comes from the rise and fall of the tide (average range is 8.4 meters) which drives turbines. The maximum power output is 240 megawatts, which is enough to supply the needs of a city of 300,000 people.

Today there is only one large tidal power station in the world, on the Rance estuary in France. Its greatest output is 240 megawatts and it produces less than one-thousandth of the energy needs of the whole country. In Britain one site, the Severn Estuary, has been studied for many years. It is thought that it might produce as much as eight per cent of the electricity demand in Britain. However, there may be other ways of producing power more cheaply.

The energy in waves is in theory very large. Scientists have shown that in the North Atlantic there is up to 100 kilowatts of energy for each meter length of wave front. Converting this energy into useful electricity and bringing it ashore would be very difficult. The energy loss on the way would be so great that only a small fraction would be left for use.

The first small wave-energy device was built in Japan about fifteen years ago by Yoshio Masuda. He used the motion of the waves rising and falling inside a vertical cylinder to push air through a turbine. The turbine was connected to an electricity generator. Hundreds of navigation buoys use his system in the Sea of Japan. The Japanese are also testing a number of other systems from several countries in a special barge, the Kaimei. Early results have been disappointing, with less than one-fiftieth of the expected output being obtained.

In some parts of the world there is a temperature difference of up to 25°C between the water's surface and some thousand meters below sea level. This difference can be used to drive a special type of turbine. Small scale experiments have shown that it might be possible to build large ocean **thermal energy** systems. This would be very expensive and there are no plans to build such systems at present.

Ever since the Earth was formed, huge amounts of heat energy have been stored in its molten core. Some of this heat is always flowing to the surface. We do not notice it as the rate of flow is very small. Hot water springs and geysers are the best known examples of **geothermal energy**. These occur where the underground rocks allow water to be heated and then return to the surface. About twenty countries use this energy for space and water heating, or to generate electricity. The total output is small, well under one per cent of world energy demand.

We could also use the heat from deep, hot dry rocks. A borehole is drilled deep into the granite. A small explosion is set off at the bottom of the hole, causing the rocks to crack. A second bore-

Deep inside the Earth the temperature rises to very high levels (the center of the Earth is estimated at 3000°C). This diagram shows how the Earth's heat can be used to produce hot water at the surface.

hole is then drilled into the cracked rocks. Water is pumped down the first hole and forced through the cracks where it is heated. It returns to the surface as very hot water. Reserves of this kind of energy are thought to be large. In Great Britain, for example, they could be equal to the energy stored in 10 billion metric tons of coal, or over one hundred times as much coal as Great Britain uses in one year.

Until now, geothermal energy has been rejected as being too expensive. However, as fossil fuels become scarcer and more expensive to extract, geothermal energy could play a vital role.

Summary

TODAY, most of the world energy needs are supplied by the fossil fuels. In one hundred years' time, coal will be the only fossil fuel left for us to use. Even if any oil or gas still remains it will be very costly and few countries will be able to afford it. Much has been said about the tar sands and oil shales in North America, but these will be much more expensive to refine and develop than crude oil is by present methods. Little of the **stored** energy is left if we go on using it at the same rate as we do today.

Fossil fuels cannot be renewed. We know that they were laid down millions of years ago as the rotting remains of plants and animals which became fossilized. This process started when the Earth was young and it cannot happen again. We have to think now of the future and look to the energy resources that will not run out. These are known as **renewable** resources, ones which mankind can go on using in any part of the world, forever.

The world is already making great use of **biomass**, meaning all forms of plant life, such as trees, which can be turned into fuels for heating and cooking, and even into liquid fuels for cars and tractors. At present, biomass makes up fifteen per cent of the world energy consumption, mostly in the developing countries. Scientists believe that biomass could be expanded as a resource by at least five times from its present level.

Hydroelectric power is already supplying six per cent of the world's energy needs. It is very costly to develop but easy and cheap to keep running. This too could be expanded by up to five times today's level.

The source of all our energy and food is the Sun itself. We have learned that all green plants convert sunlight into food by means of **photosynthesis.** It has been calculated that one hour of solar energy hitting the Earth's surface provides the same amount of energy as the world uses in one year. Some countries are already beginning to use direct sunlight to make solar heating work for them at a reasonable cost.

For the world to solve its energy problems great changes in our ways of living and working will have to take place. By the year 2000 many of the new ideas which we have talked about in this book will be working well and cheaply. As oil gets more and more expensive the new, **alternative** forms of energy will seem quite cheap.

The world has come to depend on machines for transport and industry which use oil in one form or another as fuel. In time, shortage and the high cost of fuel for these purposes will force us to develop alternative fuels. The Industrial Revolution happened as the result of coal and steam power. This was followed by the development of the internal combustion engine and the use of petroleum as a fuel. Now we take the car and its gasoline tank for granted, but this too will have to change and a new industrial revolution will take place.

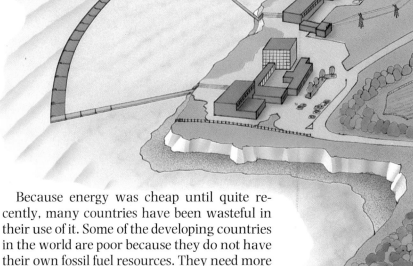

Because energy was cheap until quite recently, many countries have been wasteful in their use of it. Some of the developing countries in the world are poor because they do not have their own fossil fuel resources. They need more energy so that they can grow more food and improve their standards of living.

Life on earth depends on achieving a balance between energy supply and the needs of the world's peoples. Continued growth in the use of fossil fuels and in world population could bring great problems. Peace between nations will be more likely if there is a fairer distribution and use of all our present and future energy resources.

A future without oil

When the oil is gone or too expensive to use, energy will come from a variety of sources, as this picture shows

1 nuclear power station
2 geothermal power station
3 hydroelectric power station
4 coal-fired power station
5 coal mine
6 coal by-products
7 glasshouses
8 tidal power
9 wind power
10 wave power
11 solar energy

Glossary

absorb: to take in.

adapt: to change to suit new conditions.

air pollution: dirtying of the air caused by, for example, smoke from chimneys and fumes from engines.

algae: a group of plants that live in water. Seaweeds are algae.

alloy: a mixture of two or more metals. Brass is an alloy of copper and zinc.

alternating current: electric current flowing first in one direction then the other, all the time.

alternative energy: any new way of producing power that does not use *fossil fuels*.

amino acid: a building unit of *proteins*. The body gets amino acids by breaking down protein in food. They are needed for growth and replacing damaged or dead cells.

ampere: the unit of electrical current.

anthracite: hard, good quality coal.

anticline: a fold, like an arch, in a layer of rock.

arc lamp: a lamp in which a bright spark is made across a gap between two *conductors* when electric current runs through them.

atmospheric engine: one partly driven by air pressure.

attract: to make something come closer. A magnet attracts iron.

autotroph: a living thing that can make food from simple substances.

biomass: the total amount of animal or plant life.

biosphere: the part of the Earth, including the air above it, underground and under the sea, where things can live.

bit: the sharp, cutting end of a drill.

bituminous: containing bitumen, which is a type of tar-like substance.

blowout: a sudden strong gush of oil from an oilfield when it is being drilled.

by-product: a material also obtained when a main product is being produced.

calorie: a unit in which the energy value of food is measured.

calorimeter: a device that measures heat energy.

capacitor: a device for storing electricity.

carbon content: the amount of carbon in a substance.

carbon dioxide: a gas found in the air. (It forms the bubbles in fizzy drinks.)

carbohydrate: an energy-giving food, such as sugar or *starch*.

carnivore: an animal that eats meat.

catalyst: a substance used to speed up a chemical change in another substance. The catalyst itself is not changed.

cellulose: the material that forms the walls of plant cells.

cereal: any kind of grain, such as rice or oats, that can be used as food.

charge: to fill with electrical energy.

chemical energy: energy stored in a substance. It can be released during a chemical reaction.

combustion: the burning process which produces heat and light. Oxygen is needed for combustion to take place.

compound: a substance made up of a number of *elements*.

compression: squeezing into a smaller volume.

condensation: the process of turning a gas into a liquid by cooling it. Also, the name for the liquid that is produced.

condense: to turn from a gas or a vapor into a liquid, by cooling.

conductor: a material in which an electric current can flow easily.

continental shelf: the shallow, sloping part of the sea-bed around land.

core: the center of something, such as a nuclear *reactor*.

crankshaft: the main driving rod of an engine. It is turned by an arm attached to it and connected to a *piston*.

cruising speed: in a machine such as a car, the speed that makes best use of the fuel used, and allows the engine to run easily.

current electricity: an electric charge that flows along a *conductor*, such as a wire.

cylinder: a solid or hollow tube with circular ends and parallel sides. In an engine it is the tube in which the *piston* moves.

decomposer: an *organism* that breaks down dead animals and plants into simple substances that can be used as food by plants.

deposit: a substance which is left behind, when water has disappeared.

derrick: the frame that holds a drill over an oilfield.

digestion: the process which living things carry out to break down food so that the *nutrients* in it can be *absorbed*.

direct current: electric current flowing in one direction only.

distillation: the process of separating a mixture of liquids by heating them until they become gases and then cooling. Each part of the mixture becomes liquid again at a different temperature and can be collected.

domain: a small area in a piece of metal, such as iron, which behaves like a tiny magnet.

double-glazing: an extra sheet of glass on the inside of a window pane. Air trapped between it and the window pane stops heat escaping.

drillstring: the long chain of pipes connected to a *bit* when drilling for oil.

duodenum: the first part of the small *intestine*.

dynamo: a machine which converts movement into electrical energy.

ecosystem: any part of the natural world that exists without support from outside.

efficiency: a measure of the energy used to do something: the less energy needed, the more efficient the person or machine.

electrolysis: producing a chemical change by running an electric current through a solution.

electromotive force: the force needed to drive an electric current in an electrical circuit. It is measured in *volts*.

electron: a particle in an atom, outside the *nucleus*, with a negative electric charge.

electron flow: the passage of *electrons* from one terminal to another along a *conductor*.

element: a simple substance that cannot be broken down further by a chemical process.

environment: the surroundings of an animal or plant.

enzyme: a *catalyst* produced in living things which controls a particular chemical reaction. *Digestion* is controlled by enzymes.

equilibrium: balance.

erosion: the wearing away of the Earth's surface by wind or moving water.

esophagus: the tube that carries food from the throat to the stomach.

evaporation: the change from a liquid into vapour, as when water turns to steam.

expand: to get larger, to open out or swell.

fats: foods used by living things to produce energy. The body uses fats to keep warm. Plants and animals store fats.

fault: a break in a layer of rock where the broken edges have been forced out of place by movements of the Earth.

feces: the remains of food left after *digestion*, which are expelled from the body.

fertility: the quality of soil that lets plants grow.

fertilizer: material added to soil to improve its *fertility*.

filament: a very fine wire that glows white-hot when an electric current flows through it.

fission: the splitting of the *nucleus* of an atom. This releases a vast amount of energy.

flat plate collector: a flat glass-topped container used to collect energy from the Sun.

food chain: a series of living things starting with a plant which is eaten by an animal which is eaten by another animal, and so on.

food web: a complex pattern formed by several *food chains*. Any member of the food web may be eaten by several other members.

fossil fuel: fuel such as oil, coal, and natural gas, made from the remains of living matter.

fractionating column: a tall tube used in *distillation*: gases cool as they rise in the tube; each *condenses* at a different level and is collected.

friction: the drag when two surfaces are rubbed together. It slows down movement and produces heat.

fusion: forming a new element by joining together light atoms such as *hydrogen*. When this happens (as in the Sun), a lot of energy is produced.

galaxy: a system of stars, planets, dust and gas forming a cluster in space.

generator: a machine for turning mechanical energy into electricity.

geologist: a scientist who studies the rocks that form the Earth's crust.

geothermal energy: energy that comes from the heat inside the Earth.

glucose: a type of simple sugar made by plants during *photosynthesis*.

grid: a network of power cables that carry electricity from power stations to wherever it is used.

heat exchanger: an arrangement of pipes carrying hot and cold liquids close together. Heat passes from the hot to the cold liquid and raises its temperature.

helium: a very stable gas lighter than air. It is the simplest *element* after *hydrogen*.

herbivore: an animal that eats plants.

heterotroph: a living thing that cannot make food from simple substances, but eats animals or plants.

hydrocarbon: a chemical compound containing hydrogen and carbon atoms.

hydroelectric: producing electricity from the power of falling water.

hydrogen: a colorless, odorless gas, lighter than air. It is the simplest *element*.

ignite: to start something burning, such as a gas.

inflammable: able to burn.

insecticide: a chemical used to kill insects.

insulate: to stop passage of heat, electricity or sound.

insulator: a material in which an electric current will not flow.

internal combustion engine: an engine in which the fuel burns inside a *cylinder*, making a small explosion which drives the *piston*.

intestine: the long tube after the stomach in which food is digested, and taken into the bloodstream.

joule: a unit in which energy and work is measured.

kidneys: a pair of organs in the body which filter waste products out of the blood.

kinetic energy: energy of movement.

light energy: energy in the form of light radiation.

lignite: brown coal, softer than ordinary coal, used as fuel.

lines of flux: lines showing the direction of a force in a *magnetic field.*

liver: a large organ in the body in which many chemical reactions take place. It also produces a fluid called 'bile' which helps in *digestion.*

lodestone: a piece of brown rock containing iron, now called *magnetite.* It acts as a magnet.

lungs: a pair of organs in the body used in breathing.

magnetic field: the space around a magnet where its force can work.

magnetism: the way in which a piece of iron can *attract* or *repel* another piece of iron or an electric *conductor* placed near it.

magnetite: an iron-bearing rock that *attracts* things made of iron by *magnetism.*

mechanical energy: energy produced by machines.

migration: movement of animals or people from one place to another. Animals often migrate to and from a place depending on the time of year.

mineral salts: minerals needed by animals and plants to stay healthy.

natural gas: gas obtained from underground deposits.

neutral: with neither positive nor negative electric charge.

neutron: a *particle* without an electric charge in the *nucleus* of an atom.

nomad: one of a group of people who wander about with livestock in search of pasture.

nuclear fission: see *fission.*

nucleus: the central part of an *atom*, containing nearly all its mass.

nutrient: a substance that an animal or plant needs to take in as food.

oasis: a desert settlement around a water supply.

ohm: the unit of electrical *resistance.*

oil pollution: oil spilt in the sea, which forms a thick layer. This kills seabirds and other living things.

optimum speed: see *cruising speed.*

organism: a living animal or plant.

Otto cycle: the working of a four-stroke engine: sucking in fuel; *compression*; explosion; pushing down the *piston*; driving out waste gas.

output: the work produced by a process that uses energy.

oxygen: one of the gases found in air and water. Living things need it for *respiration.*

ozone: a form of *oxygen* which *absorbs* ultraviolet rays from the Sun. A layer in the atmosphere prevents these rays reaching Earth and destroying life.

particle: a very small piece of solid material.

peat: a layer of decayed plants found on the surface of the ground in boggy areas, and used as fuel.

phloem: tiny tube-like cells in plants which carry food down from the leaves.

phosphate: one of a group of chemicals used as a *fertilizer* to make plants grow bigger.

photosynthesis: the process by which green plants use energy from sunlight to make their food from water and *carbon dioxide.*

physiologist: a scientist who studies the working of living things.

piston: a piece of metal which fits inside a *cylinder* and moves up and down. It may act as a pump or be pushed by a liquid or gas to produce power.

pole: one of the points at each end of a magnet.

pollutant: a substance that causes *pollution.*

pollution: the act of making the *environment* dirty and harmful with rubbish, fumes and waste chemicals.

polypropylene: a strong, hard-wearing plastic.

polyurethane: plastic used to make paint and varnish, and foam plastics for padding.

potential energy: energy stored in an object because of its position.

protein: body-building food, found in meat, fish, eggs, etc.

proton: a *particle* with a positive electric charge in the *nucleus* of an atom.

radioactivity: the breaking down of atoms in some heavy *elements.* Dangerous rays are given out by the reaction.

reactor: the container in a nuclear power plant in which atoms are split to produce great heat.

rectum: the final part of the large *intestine*, from which undigested food leaves the body.

recycle: to re-use old products to make new ones.

refinery: a place where raw materials such as oil are cleaned and divided into different grades.

renewable: able to be made new over and over again.

repel: to push something away.

resistance: the way in which a material slows down an electric current flowing through it. Good *conductors* have low resistance.

resource: something which supports us; a supply of energy.

respiration: in animals and plants, taking in *oxygen* from the air and giving out *carbon dioxide*, or breathing. Also, the way in which living things use oxygen to break down their food to get energy.

rotor: the turning part of a motor or *generator.*

roughage: food that is not broken down by the body during digestion. It helps to make the *feces* soft and helps to avoid constipation.

saliva: the fluid in your mouth that is mixed with food so that you can swallow it. It also contains an *enzyme* which helps to *digest starch.*

self-sufficiency: living without depending on other people or other communities for food.

semiconductor: a material in which only part of an electric current can flow.

settlement: a place where people make homes and live permanently.

solar cell: a unit containing silicon which changes radiation from the Sun into electrical energy.

solar energy: energy radiated by the Sun.

solenoid: a wire coil carrying an electric current which behaves like a magnet.

starch: a *carbohydrate* food material stored by plants. Starchy foods include bread and potatoes.

static electricity: an electric charge that stays in one place. It builds up on materials which are not *conductors.*

stored energy: energy contained in fuels which can be turned into other forms of energy.

stroke: one movement up and down of a *piston.*

sulfur dioxide: a gas formed when sulfur burns and a *pollutant* found in smoke.

tension: a strain produced by pulling or stretching.

thermal energy: energy that comes from heat.

town gas: coal gas, released when coal is heated.

transformer: a device for changing the voltage, or strength, of an electric current.

transistor: a small electric valve through which current flows in one direction only at a controlled rate, through solid matter. It can be used to increase the power of an electric signal.

tropical rain forest: large forest areas growing in hot countries which have a high rainfall.

turbine: a wheel of curved blades inside a tube, which is turned by water or steam. The energy is used to make electricity.

uranium: a heavy grey *radioactive* metal used as fuel in a nuclear *reactor.*

vacuum: an empty space with no air or any other matter in it.

vitamins: various substances needed by the body in very small amounts in order to stay healthy.

volt: the unit in which *electromotive force* is measured.

whipstock: a wedge used to change the direction of a drill.

water pollution: the pumping of sewage or waste chemicals into water. This can make it unfit to drink and kill animals and plants.

xylem: tiny tube-like cells in plants which help in support. They carry water and minerals up from the roots.

Index

adaptation 12
air 22; pressure 27; pollution 52, 53
algae 53
alternator 40
alternating current 40
amber 30
amino acid 19, 21
Ampère, André 37
ampere 37
anticline 44
arc lamp 32, 33
astronaut 25
atmospheric engine 27
atom 34; nucleus 34; proton 34; electron 34
autotroph 16

Ball of Aeolus 26
barley 23
battery 32, 35
Bedouin 13
beef cattle 19
bicycle 15
biomass 60
biosphere 16
bituminous coal 42
blowout 45
body, human 20, 21
borehole 59
bushmen 12

calorie 18
capacitor 31
carbohydrate 18, 19, 21
carbon content 42
carbon dioxide 21, 22
carnivore 16, 17
catalyst 49
Cavendish laboratory 50
cell 36
cellulose 22
cereal crop 23
chain reaction 50, 51
chemical energy 23
chemical waste 52, 53
China 5
civilization 14
coal 4, 7, 42, 43, 60; anthracite 42; by-product 43; carbon 42; gas 42; lignite 42; mining 26, 42, 43
compression 24
computer 29
condensation 20, 26, 49, 57
conductor 34, 36
conservation 54, 55; forest 54; fuel 55
continental shelf 46
copper 35, 36, 38
crankshaft 28, 29
crude oil 48, 49
cruising speed 15
cylinder 26, 28, 29, 59

dam 11, 58
Davy, Sir Humphry 32
decomposers 17
DDT 53
Deptford 40
derrick 44
desert 10, 12
developing countries 4, 5
diet 17, 18, 19
digestion 20, 21
direct current 40

distillation 49, 57
double glazing 55
drilling 44, 45, 46
drillstring 45, 46
duodenum 20
dynamo 33, 39

early man 14
ecosystem 16, 17; pond 16
Edison, Thomas 33, 40
Einstein, Albert 50
electricity 6, 7, 30, 31, 32, 33, 34, 35, 36, 37, 38, 39, 40, 41, 56, 57, 58; amber 30; amp 37; battery 32, 35, 36; capacitor 31; cell 36; charge 35; conductor 34, 35; current 31, 32, 33, 35, 36, 38, 39; dynamo 33, 39; electric car 29; electric shock 31; electromotive force 36; filament 33; generator 33, 39; hydroelectric power 6, 41, 58; Leyden jar 31; insulator 34, 35; lightning 31; magnetic effect 32; motor 39; power station 6, 7, 40, 41; pressure 36; resistance 36, 37; static 30, 35; transformer 41; voltage 35, 36, 37, 41
electromagnet 39
electron 34, 36; flow 36
electrolysis 32
element 8, 15, 18, 19, 21, 22, 27
energy 4, 5, 6, 7, 8, 9, 11, 38, 39, 44, 45, 50, 51, 58, 59, 60, 61; alternative 60; chemical 23; conservation 5, 55; crisis 5; electrical 6, 7; from tides 58, 59; from water 58, 59; kinetic 6, 11; light 23; mechanical 11, 28, 29; oil 44, 45; potential 6, 11; shortage 4, 5; solar 2, 56, 57, 60; sources 4, 5, 6, 7, 8, 9, 11, 60; stored 60; thermal 59
engine 26, 28, 29; atmospheric 27; cylinder 26, 28, 29; four-stroke 28, 29; Otto cycle 28, 29; piston 26, 28, 29; steam 26, 27, 28; valve 29
environment 12, 13, 29, 52, 53, 58
enzyme 20
erosion 12
Eskimo 12, 13
esophagus 20
evaporation 9, 10, 54
expansion 26

Faraday, Michael 32, 33, 39, 40
fat 8, 18, 19
fault 44
du Fay, Charles 30
feces 21
Ferranti 40
fertilizer 10, 52
fertility 23
fission 8, 50, 51
flat plate collector 56
food 4, 16, 17, 18, 19, 20, 21, 23; amino acid 19; calorie 18; carbohydrate 18, 19; chain 16, 17, 19; digestion 20, 21; mineral salt 19; nutrient 18, 19; protein 19; roughage 19; vitamin 19; web 16, 17
force 24, 25; in equilibrium 24; of gravity 24, 25; natural 24, 25
forest 12, 54
fossil fuel 4, 9, 42, 43, 44, 55, 56, 60
fractionating 49
friction 24
Franklin, Benjamin 31
fuel 4, 5, 15, 18, 19, 28, 29, 54, 55, 60

fusion 8, 51

galaxy 8
Galvani, Luigi 32
gas 4, 8, 29, 43, 60
generator 33, 39
geologist 44
geothermal energy 59
Gilbert, Sir William 30, 38
glucose 22
gravity 24, 25
greenhouse effect 9
grid 41

hard water 10
heat exchanger 51
helium 8
Henry, Joseph 38
herbivore 16
Hero of Alexandria 26
heterotroph 16
hydrocarbon 49
hydroelectric power 6, 11, 41, 58, 60
hydroelectric power station 6, 11, 58
hydrogen 8, 10, 34

Industrial Revolution 27
insecticide 53
insulation 9, 55
internal combustion engine 28, 29, 60
intestine 19, 20, 21; duodenum 20; large 20; rectum 21; small 20, 21

joule 18

Kalahari desert 12
kidney 21
kinetic energy 6, 11

Lapps 13
Leyden jar 31
light 9, 30
lightning 31
liver 21
lodestone 32, 38
lung

magnetism 32, 38; domain 38; electromagnet 39; lines of flux 38; magnetic field 38; magnetic force 38; magnetite 38; solenoid 38
market 14, 15
Masuda, Yoshio 59
migration 13
Milky Way 8
mineral salt 19, 22
mines 26, 27
motor car 28, 29
van Musschenbroek, Pieter 31

natural forces 24, 25
neutron 34
Newcomen, Thomas 27
Newton, Sir Isaac 6
nichrome wire 16
nomad 13
nuclear power 5, 7, 50, 51; chain reaction 50, 51; fission 50, 51; heat exchanger 51; radio-activity 50, 51; reactor 50, 51; station 7, 50, 51; uranium 7, 50, 51
nucleus 34
nutrient 18, 19

oasis 10
Odeillo 56
Oersted, Hans 32, 38
ohm 37
Ohm, Georg Simon, 37

Ohm's Law 37
oil 4, 5, 28, 29, 44, 45, 46, 47, 48, 49, 60; by-products 44, 49; crude 48, 49; drilling 44, 45, 46, 47; drilling rig 46; drill-string 46; exploration 46, 47; prospecting 44; pollution 52; production platform 46, 47; offshore rigs 46, 47; refinery 49; refining 49; terminal 46; wells 46, 47; whipstock 46
orbit 25
Otto cycle 28, 29
Otto, Nikolaus August 28, 29
output 4
oxygen 3, 21, 22
ozone 53

Parsons, Sir Charles 40
particles 50
peat 42
phloem 22
phosphate 52
photosynthesis 22, 60
piston 26, 28, 29
plant 16, 17, 22, 23
plastics 49
pole 38
pollution 5, 29, 42, 52, 53
polypropylene 49
polyurethane 49
population 4, 12, 14, 15, 60
potential energy 6, 11
power 26, 27, 28, 33, 50, 51, 58
power stations 6, 7, 27
production platform 46, 47
protein 19, 20, 23
proton 34
pump 26, 27

radioactive waste 53
radioactivity 50, 51
rain 10, 54
rain forest 54
Rance estuary, tidal power 58
recycling 54
renewable resources 5
resistance 36, 37
resistor 37
resources 4, 13, 60
respiration 22
rice 23
roughage 19
Royal Institution 32
Rutherford, Ernest 50

saliva 20
Savery's engine 27
Savery, Thomas 27
scurvy 19
sea 10
self-sufficiency 14
semiconductor 35
Severn Estuary 59
snow 10
soft water 10
soil 22
solar energy 5, 8, 9, 56, 57, 60; cells 57; pond 9; uses 57
solar radiation 9, 56
solenoid 38
starch 20
steam 26, 27, 28; engine 26, 27; power from 26, 27; turbine 27, 50
stomach 20
Sturgeon, William 38
sugar 20, 21, 22
sulfur dioxide 52
sun 8, 9, 10, 22, 56, 60
sunlight 16, 17, 22
Swan, Sir Joseph 33

Tardier, Jean 43
tension 24

Thales 30
thermal energy 59
tidal power station 58, 59
transformer 41
transistor 35
transport 15, 28, 29
travel 15
tropical rain forest 54
turbine 11, 27, 40, 50, 58; steam 27, 40; water 11, 58

uranium 7, 50, 51

vacuum 26
valve 29
vitamins 19
Volta, Alessandro 32, 36
Volta's pile 32
volt 35, 36, 37
voltmeter 37

waste 21, 52, 54
water 5, 10, 11, 58, 59; force 24; fall 11; pollution 52; power 5, 11, 58, 59; supply 10; use of 10, 11, 58, 59; vapor 10; wheel 11
wavepower 5, 58, 59
weight 25
weightlessness 25
whipstock 46
wind 24, 57; force 24
windmill 57

xylem 22

zero gravity 25